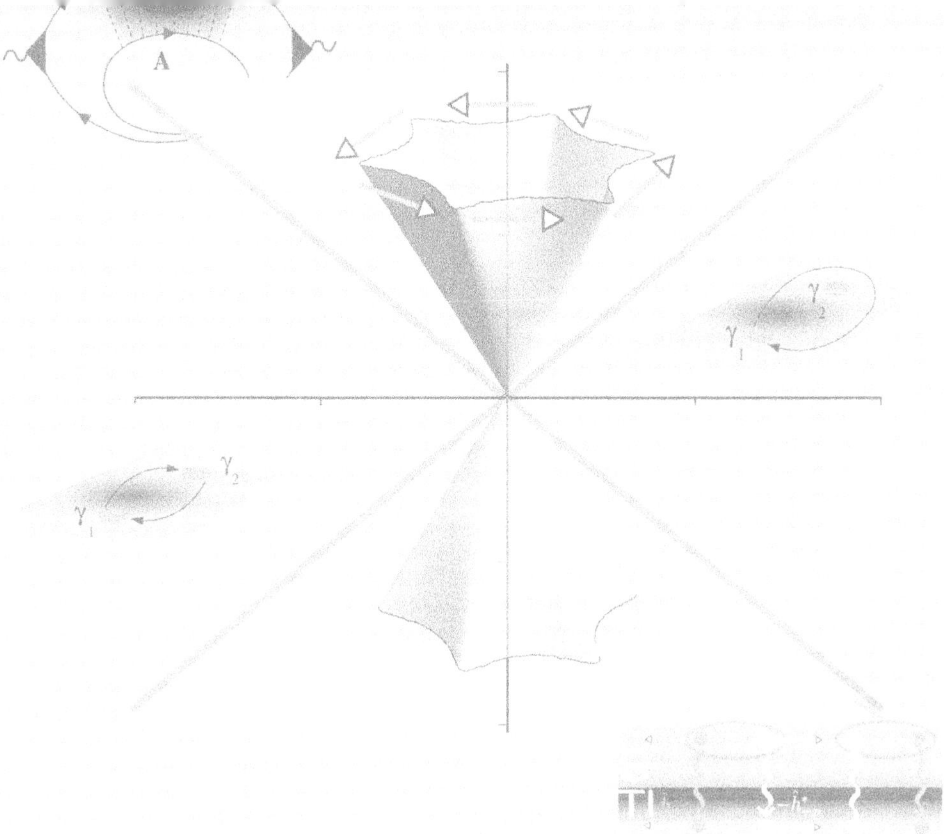

Topological Insulators

Topological Insulators

The Physics of Spin Helicity in Quantum Transport

Grigory Tkachov

Pan Stanford Publishing

Published by

Pan Stanford Publishing Pte. Ltd.
Penthouse Level, Suntec Tower 3
8 Temasek Boulevard
Singapore 038988

Email: editorial@panstanford.com
Web: www.panstanford.com

British Library Cataloguing-in-Publication Data
A catalogue record for this book is available from the British Library.

Topological Insulators: The Physics of Spin Helicity in Quantum Transport

ISBN 978-981-4613-25-5 (Hardcover)
ISBN 978-981-4613-26-2 (eBook)

Printed in the USA

Contents

Preface

This book came into being under the influence of dynamic developments that have occurred in condensed matter physics after the recent discovery of a new class of electronic materials called topological insulators. From the viewpoint of the textbook classification of solids, topological insulators are rather peculiar materials. In terms of electric conductivity they are neither ordinary insulators nor conventional metals. In a topological insulator, the character of electron transport varies from insulating in the bulk of the material to metallic at its surface. Such an atypical character of electric conduction stems from unusual electronic states that, for topological reasons, appear at the surface of the material. In the two-dimensional geometry, the topological states are bound to the edges of the system. The surface or edge states are protected by time-reversal symmetry and characterized by nontrivial topological quantum numbers. These novel properties distinguish topological insulators from ordinary band insulators and metals.

Far from being an abstract construction, topological materials can be synthesized by means of modern crystal growth techniques. The first experimentally discovered topological insulator was a two-dimensional quantum spin-Hall system created on the basis of HgTe quantum-well structures. This system exhibits a new type of the Hall effect—the quantum spin-Hall effect—that occurs in a zero external magnetic field and is observable through the conducting edge states forming a Kramers' doublet. Other subsequently discovered members of the family of topological materials include the semiconducting alloy $Bi_{1-x}Sb_x$, tetradymite compounds, e.g., Bi_2Se_3, Bi_2Te_3, and Sb_2Te_3, strained three-dimensional layers of HgTe, just to name a few.

One of the goals of this book is to expose the topological origin of the edge and surface states in two- and three-dimensional systems. To that end, several theoretical models of the topological states are introduced and discussed in detail. These models admit analytical solutions, vividly illustrating key properties of the edge and surface states such as their Dirac-like spectrum and spin helicity. Remarkably, the helicity of the edge and surface states is preserved in collisions with impurities and other crystal or sample defects, due to which topological insulators have been considered as platforms for many intriguing applications, ranging from spintronics to topological quantum information processing.

The realization of the application potential of topological insulators requires a comprehensive and deep understanding of transport processes in these novel materials. The book is therefore intended to give an insight into some representative transport phenomena in which the key role belongs to the spin helicity. These include the quantum spin-Hall effect, nonlocal edge transport, backscattering of helical edge and surface states, weak antilocalization, unconventional triplet p-wave superconductivity, topological bound states, and emergent Majorana fermions in Josephson junctions as well as superconducting Klein tunneling.

The book is self-contained to the extent that its material can be integrated into the courses of condensed matter physics for undergraduate or graduate students. The main text is complemented by end-of-chapter problems. Methodologically, this book is an attempt to present key ideas and concepts of the physics of topological insulators to a wide spectrum of physicists, while introducing and revisiting important methods of condensed matter and theoretical physics. Selected topics of the theory of electron localization, unconventional superconductivity and high-magnetic field effects would be of interest to active researchers as well.

The author is grateful to E. M. Hankiewicz, L. W. Molenkamp, B. Trauzettel, H. Buhmann, A. Pimenov, K. Richter, P. Recher, M. Hentschel, and many other collaborators for stimulating discussions.

Grigory Tkachov
Wuerzburg
September 2015

Acknowledgements

Part of the work on this book was conducted under the grant No TK60/1-1 of the German Research Foundation (DFG). The author gratefully acknowledges the support of the German Research Foundation.

Chapter 1

Introducing Topological Insulators: Mind the Time Reversal

This chapter is intended to acquaint the reader with a general notion of topological insulators. The content of the chapter captures the basic ideas and concepts of the pioneering studies of the two-dimensional topological insulators (see Refs. [1–7]) and their three-dimensional counterparts (see Refs. [8–12]). We shall see that the emergence of topological insulators is a vivid example of the importance of fundamental symmetries in physics. In our case, it is the symmetry with respect to the reversal of the direction of time. We, therefore, begin with a discussion of general ideas about time-reversal invariance and its breaking, which will lead us step-by-step to the notion of topological insulators.

1.1 Time-Reversal Symmetry, Anti-Unitary $\mathbb{T}$ Operation, and Kramers' Theorem

Time-reversal symmetry (TRS) means the invariance of physical properties of a system under the change of the time arrow,

$$t \mapsto -t. \tag{1.1}$$

Topological Insulators: The Physics of Spin Helicity in Quantum Transport
Grigory Tkachov

ISBN 978-981-4613-25-5 (Hardcover), 978-981-4613-26-2 (eBook)
www.panstanford.com

For a particle with linear momentum $\boldsymbol{p}$ and spin $\boldsymbol{s}$ located at position $\boldsymbol{r}$, the TR implies the reversal of the directions of both $\boldsymbol{p}$ and $\boldsymbol{s}$ at the same position:

$$\boldsymbol{r} \mapsto \boldsymbol{r}, \qquad \boldsymbol{p} \mapsto -\boldsymbol{p}, \qquad \boldsymbol{s} \mapsto -\boldsymbol{s}. \tag{1.2}$$

The system is said to be TR invariant if under transformations (1.2) its Hamiltonian, $\hat{h}$, does not change. This is obviously true if the Hamiltonian is an even function of the momentum and independent of the spin, for example, the Hamiltonian of a free non-relativistic particle. A more interesting example of the TRS is a Hamiltonian that is odd in both momentum and spin, for example,

$$\hat{h} = v\boldsymbol{\sigma} \cdot \boldsymbol{p} = \frac{2v}{\hbar}\boldsymbol{s} \cdot \boldsymbol{p}, \qquad \boldsymbol{s} = \frac{\hbar}{2}\boldsymbol{\sigma}, \tag{1.3}$$

where $\hbar = h/(2\pi)$ is Plank's constant, and the vector $\boldsymbol{\sigma} = (\sigma_x, \sigma_y, \sigma_z)$ consists of 2×2 Pauli matrices

$$\sigma_x = \begin{bmatrix} 0 & 1 \\ 1 & 0 \end{bmatrix}, \qquad \sigma_y = \begin{bmatrix} 0 & -i \\ i & 0 \end{bmatrix}, \qquad \sigma_z = \begin{bmatrix} 1 & 0 \\ 0 & -1 \end{bmatrix}. \tag{1.4}$$

Hamiltonian (1.3) describes a massles spin-1/2 particle moving with the velocity, v. More generally, the TRS holds for any intrinsically nonmagnetic system in the absence of an external magnetic field.

Irrespective of the details of the system, the TRS can be formally defined as the invariance of the Hamiltonian under a certain operation, $\mathbb{T}$:

$$\mathbb{T}\hat{h}\mathbb{T}^{-1} = \hat{h} \qquad \Longleftrightarrow \qquad \mathbb{T}\hat{h} = \hat{h}\mathbb{T}, \tag{1.5}$$

under conditions

$$\mathbb{T}\boldsymbol{r}\mathbb{T}^{-1} = \boldsymbol{r}, \tag{1.6}$$

$$\mathbb{T}\boldsymbol{p}\mathbb{T}^{-1} = -\boldsymbol{p}, \tag{1.7}$$

$$\mathbb{T}\boldsymbol{s}\mathbb{T}^{-1} = -\boldsymbol{s}, \tag{1.8}$$

where $\mathbb{T}^{-1}$ is the inverse operation. In order to find the representation for $\mathbb{T}$ we notice that Eqs. (1.6) and (1.7) with $\boldsymbol{p} = -i\hbar\nabla$ imply that the operator $\mathbb{T}$ involves complex conjugation, C, that is, $\mathbb{T} = UC$, where U must be a unitary matrix to preserve the norm of the state. The only possible representation compatible with Eq. (1.8) is $U = \pm i\sigma_y$. Without the loss of generality we can define the TR operation as

$$\mathbb{T} = i\sigma_y C. \tag{1.9}$$

This is an example of anti-unitary operators introduced to physics by Wigner [13]. Generally, an operator $\mathbb{A}$ is called anti-unitary if it maps the Hilbert space on itself, preserving the norm, and the scalar product of any two states $\mathbb{A}|a\rangle$ and $\mathbb{A}|b\rangle$ is given by

$$\langle a|\mathbb{A}^\dagger\mathbb{A}|b\rangle = \langle a|b\rangle^* = \langle b|a\rangle. \tag{1.10}$$

This differs from the unitary case by the extra complex conjugation denoted above by the asterisk (whereas, $\dagger$ means Hermitian conjugation). In addition to Eq. (1.10), the $\mathbb{T}$ operation has the property

$$\mathbb{T}^2 = -1 \qquad \Longleftrightarrow \qquad \mathbb{T}^{-1} = -\mathbb{T}. \tag{1.11}$$

We are now prepared to prove a theorem that has important implications for TR-invariant quantum mechanical systems. It is called Kramers' theorem. It states that each energy level of a TR-invariant system with half-integer spin is at least double degenerate. Let us consider two states $|\pm\rangle$ related to each other by the TR operation:

$$|+\rangle = \mathbb{T}|-\rangle. \tag{1.12}$$

If $\mathbb{T}$ is a symmetry of the Hamiltonian, the two states $|\pm\rangle$ correspond to the same energy. We need to prove that $|\pm\rangle$ are two independent states, which is the case if they are orthogonal to each other, that is,

$$\langle +|-\rangle = 0. \tag{1.13}$$

Using Eqs. (1.10), (1.11), and (1.12), we can rewrite the product $\langle +|-\rangle$ as

$$\langle +|-\rangle = \langle -|\mathbb{T}^\dagger\mathbb{T}^{-1}|+\rangle = -\langle -|\mathbb{T}^\dagger\mathbb{T}|+\rangle = -\langle -|+\rangle^* = -\langle +|-\rangle. \tag{1.14}$$

This necessarily leads to the conclusion that $\langle +|-\rangle$ vanishes identically, proving this theorem. As we shall see, many properties of topological insulators can be understood as a consequence of fundamental Kramers' theorem.

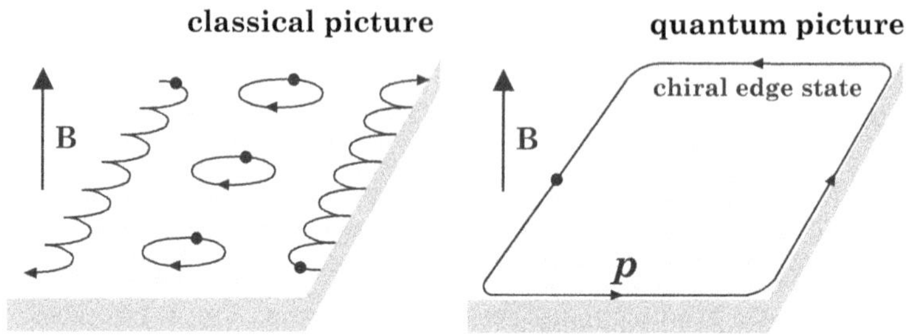

Figure 1.1 Two-dimensional electron system in a perpendicular external magnetic field $\boldsymbol{B}$. The formation of a chiral extended edge state manifests broken time-reversal symmetry.

1.2 Broken Time-Reversal Symmetry, Chiral Edge States and Quantum Hall Effect

Breaking the TRS can have dramatic consequences. An important (and very relevant to our subject) example of the TRS breaking is illustrated in Fig. 1.1. The figure shows a two-dimensional (2D) electron system subject to a perpendicular external magnetic field $\boldsymbol{B}$. In the classical picture, the magnetic field forces a charged particle to move along curved trajectories that can be divided into two groups. One of them includes closed orbits appearing in the interior (bulk) of the sample away from its edges. The other group consists of open electron trajectories that hit repeatedly the sample edge. The open orbits accommodate specific electronic states that can skip all the way around the sample perimeter. These skipping trajectories indicate the formation of propagating edge states in a magnetic field.

In a strong magnetic field, when the periodic carrier motion is quantized, a skipping trajectory transforms into a quasi-one-dimensional edge channel encircling the interior of the system [14, 15]. The direction of the edge-state momentum $\boldsymbol{p}$ is tied to the orientation of the magnetic field. Such directiveness of the edge state protects it against disorder which is inevitably present, to a certain degree, in any realistic system. At low temperatures, the main source of disorder is random fluctuations of the electrostatic background that tend to induce elastic momentum scattering with $\boldsymbol{p} \mapsto -\boldsymbol{p}$. However, for the edge state such backscattering processes

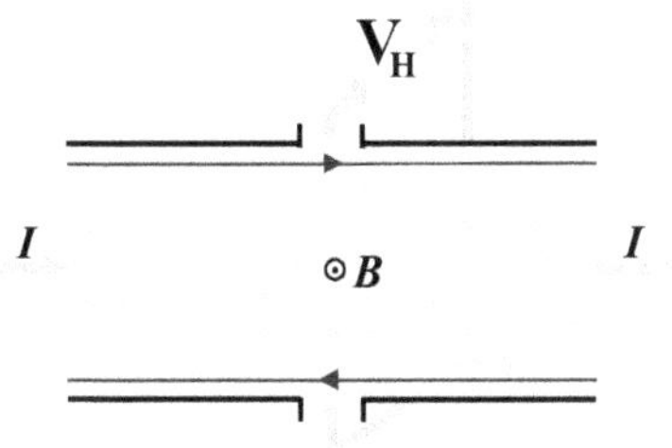

Figure 1.2 Schematic of a quantum Hall state with chiral channels at the sample edges.

are rigorously forbidden in view of the absence of the available counter-propagating state. The robust directiveness of the edge states is called chirality. It is a rather nontrivial manifestation of the broken TRS that explains the dissipationless character of the well-known macroscopic quantum phenomenon, the quantum Hall effect [16].

In the Hall effect a voltage difference, V_H, in a direction transverse to the electric current, I, is produced by applying an external magnetic field (see also Fig. 1.2). In strong magnetic fields the Hall conductivity $\sigma_H = I/V_H$ exhibits quantization in integer units of the quantum $2e^2/h$ [16],

$$\sigma_H = \frac{2e^2}{h} n, \tag{1.15}$$

where the factor of 2 is the spin degeneracy, e is the electron charge and $n = \pm 1, \pm 2, \ldots$ is integer, while the conductivity in the direction of the current (the longitudinal conductivity) is vanishingly small. In this quantum regime the electric current produces no resistive heating, as the directions of the current flow and electric field are perpendicular to each other. Such dissipationless quantum Hall transport is naturally attributed to the chiral edge states that are capable of carrying the electric charge without scattering and, hence, without generating resistance along the edge. The necessary condition for that is the absence of the mobile bulk carriers. Therefore, the edge modes of a quantum Hall system must occur in the energy gap between the bulk bands, as sketched in Fig. 1.3. The quantum Hall effect is robust against continuous deformations of the

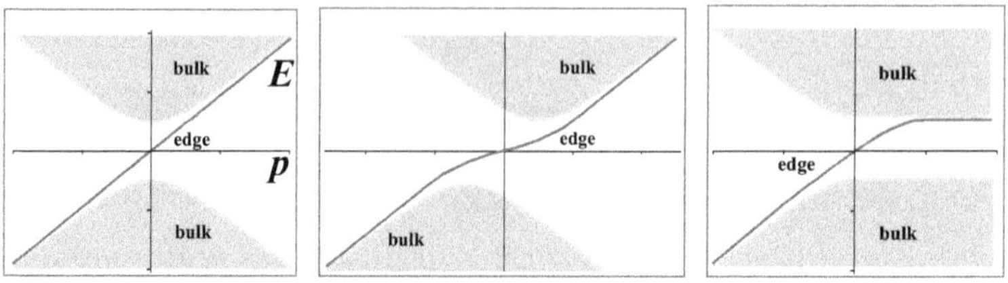

Figure 1.3 Examples of the band structure of a quantum Hall system with a single gapless edge mode. The three band structure types correspond to the same quantum Hall state as they are connected by continuous deformations.

band structure as long as the bulk band gap and the edge modes remain intact.

1.3 From Chiral to Helical: Topological Insulators in two Dimensions

The existence of the robust edge states is also central to the physics of the recently discovered 2D topological insulators (2DTIs) [1–7]. However, unlike the quantum Hall systems discussed above, the 2DTIs feature edge states in a zero magnetic field, that is without breaking the TRS. This becomes possible due to intrinsic spin–orbit coupling effects in TI materials. In the simplest picture the spin–orbit coupling can be viewed as an intrinsic effective magnetic field $\boldsymbol{B}_{\text{eff}}$ that points in the opposite directions for the up- and down-spin species, as sketched in Fig. 1.4. The two spin subsystems can be viewed as two copies of a quantum Hall insulator with a gapless edge mode each, resulting in a *pair* of edge states in total in a zero magnetic field. The defining feature of the edge states in 2DTIs is the locking between the spin and momentum directions, also referred to as *helicity*. Such helical edge states form a Kramers doublet with the energy-momentum dispersions crossing in the bulk band gap, as depicted in Fig. 1.5.

Although in 2DTIs the edge states are available in both directions, the scattering between them is forbidden for any single-particle scattering potential preserving the TRS. In order to understand this in more detail, we notice that the helical edge states transform into

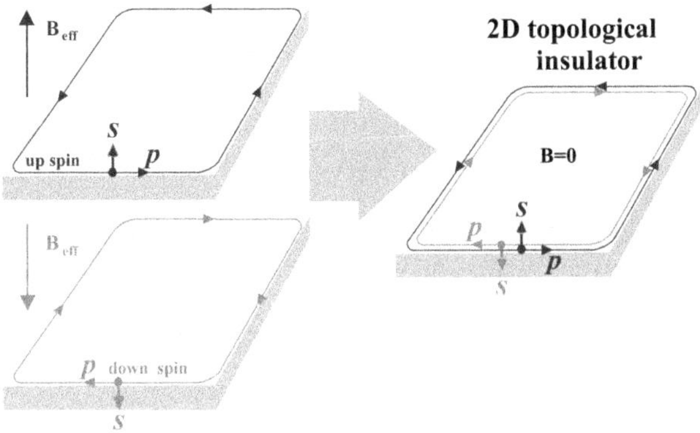

Figure 1.4 Schematic of helical edge states in a two-dimensional topological insulator.

each other upon the simultaneous reversal of the momentum and spin. Since this implies the reversal of time [cf. Eq. (1.1)], the two edge states are related by the TR operation: $|+\rangle = \mathbb{T}|-\rangle$, where $|\pm\rangle$ refer to the right- and left-moving states, respectively (see also Fig. 1.6). The absence of scattering between the edge states means the vanishing of the matrix element

$$\langle +|\hat{V}|-\rangle = 0, \tag{1.16}$$

where $\hat{V}$ is the scattering potential. We can prove this by analogy with Kramers's theorem. Let us consider a TR-invariant potential $\hat{V}$ satisfying

$$\mathbb{T}\hat{V}\mathbb{T}^{-1} = \hat{V} \quad \Longleftrightarrow \quad \mathbb{T}\hat{V} = \hat{V}\mathbb{T}. \tag{1.17}$$

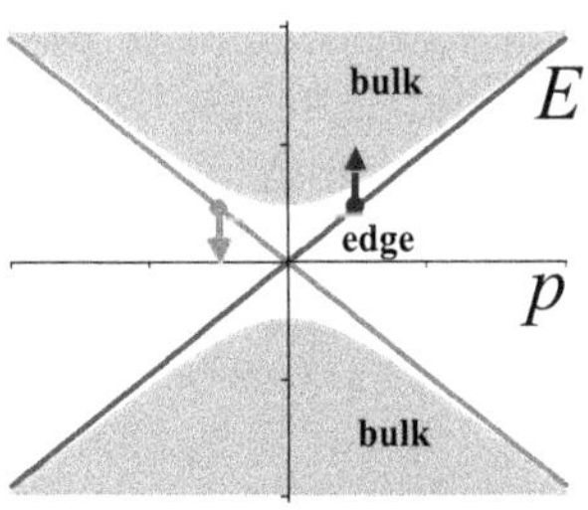

Figure 1.5 Schematic band structure of a 2DTI with helical edge states in the gap between the bulk conduction and valence bands.

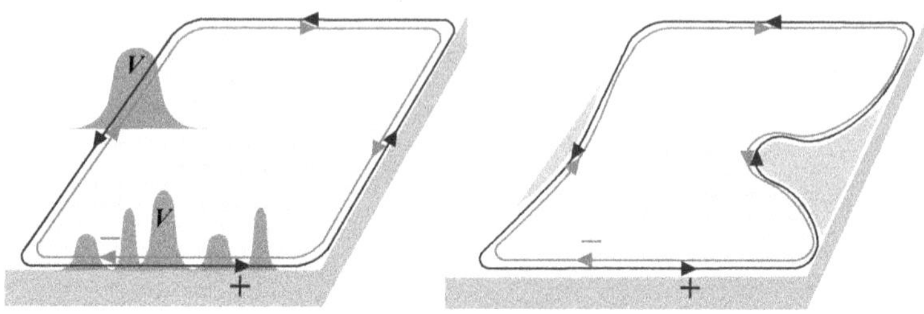

Figure 1.6 Illustration of robust helical edge states in an irregular potential landscape.

Using Eqs. (1.11), (1.12), (1.10) and (1.17), we can recast the scattering matrix element as follows:

$$\begin{aligned}\langle+|\hat{V}|-\rangle &= \langle-|\mathbb{T}^{\dagger}\hat{V}\mathbb{T}^{-1}|+\rangle = -\langle-|\mathbb{T}^{\dagger}\hat{V}\mathbb{T}|+\rangle = -\langle-|\mathbb{T}^{\dagger}\mathbb{T}\hat{V}|+\rangle \\ &= -\langle-|\hat{V}|+\rangle^{*} = -\langle+|\hat{V}^{\dagger}|-\rangle. \end{aligned} \tag{1.18}$$

Clearly, for a physical hermitian perturbation ($\hat{V} = \hat{V}^{\dagger}$) the matrix element (1.18) is identically zero.

As a consequence of Eq. (1.16), the irregularities of the potential landscape of the sample cannot reverse the propagation directions of the helical edge states [see Fig. 1.6(left)]. Even though the shape of their trajectories may change near certain obstacles [see Fig. 1.6(right)], they still remain the "time-reversed partners" in the sense of Kramers' theorem, and their global conduction capability is just the same as in the ideal system. This falls into the general notion of topological objects as systems whose global characteristics are invariant under continuous local deformations. Further, topological aspects of TIs, such as the band inversion effect and topological quantum numbers, will be discussed in the next chapter.

In order to better appreciate the topological robustness of the helical edge states, it is instructive to compare them with conventional one-dimensional conductors in the presence of potential disorder. In the conventional systems, all propagating states become localized due to backscattering off the random potential, a phenomenon known as *Anderson localization* [17]. It occurs for arbitrary weak disorder provided that the system is large enough. Equation (1.16) implies the absence of Anderson localization in TIs, which distinguishes them qualitatively from ordinary conductors.

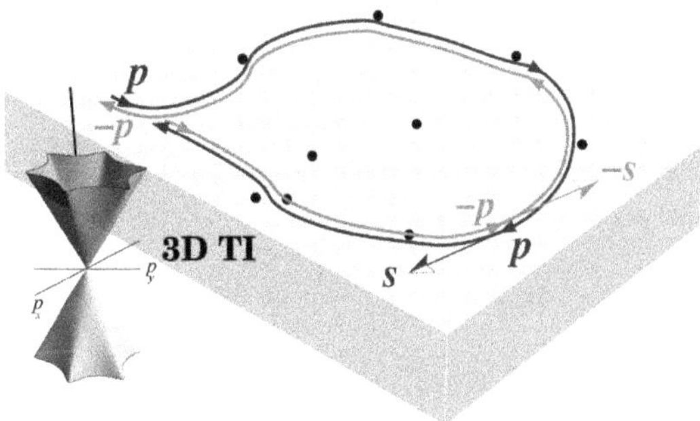

Figure 1.7 Schematic of Dirac-like propagating states on the surface of a three-dimensional topological insulator (3DTI).

Owing to the spin-momentum locking in the edge channels, the electronic state of the 2DTI is also called the quantum spin-Hall state. It was predicted theoretically in Refs. [1, 2] for graphene with spin–orbit coupling and in Refs. [3, 4] for semiconductor quantum wells. Experimentally, the quantum spin-Hall state has been observed and investigated, for the first time, in HgTe quantum-well structures [5–7].

1.4 Three-Dimensional Topological Insulators

The higher dimensional analogues of the system considered above are the three-dimensional topological insulators (3DTIs) [18, 19]. They support protected electronic states on the surface of a bulk material (see Fig. 1.7), while the inner part of the system behaves as an insulator. A surface carrier has a freedom to move in two dimensions, but, similar to the edge states, its spin $\boldsymbol{s} = \frac{\hbar}{2}\boldsymbol{\sigma}$ is locked in the direction of the momentum $\boldsymbol{p}$. The simplest Hamiltonian describing the surface states with such spin-momentum locking is $\hat{h} = v\boldsymbol{\sigma} \cdot \boldsymbol{p}$, as introduced earlier in Eq. (1.3), where the particle momentum $\boldsymbol{p}$ should be treated as a two-dimensional vector on the surface:

$$\boldsymbol{p} = (p_x, p_y, 0). \tag{1.19}$$

The surface Hamiltonian can then be written as a 2 × 2 matrix:

$$\hat{h} = v(\sigma_x p_x + \sigma_y p_y) = v \begin{bmatrix} 0 & p_x - ip_y \\ p_x + ip_y & 0 \end{bmatrix}. \quad (1.20)$$

The energy $E_{\boldsymbol{p}}$ of a surface state with momentum $\boldsymbol{p}$ is obtained by solving the eigenvalue equation:

$$v\boldsymbol{\sigma} \cdot \boldsymbol{p}\psi = E_{\boldsymbol{p}}\psi, \qquad \psi = \begin{bmatrix} \psi_{\uparrow} \\ \psi_{\downarrow} \end{bmatrix}, \quad (1.21)$$

where the wave function ψ is a spinor with two components $\psi_{\uparrow}$ and $\psi_{\downarrow}$. Solving Eq. (1.21) for $E_{\boldsymbol{p}}$ yields

$$E_{\boldsymbol{p}} = \pm v|\boldsymbol{p}| \equiv \pm v\sqrt{p_x^2 + p_y^2}. \quad (1.22)$$

The dispersion relation $E_{\boldsymbol{p}}$ has the form of a cone, with the signs ± corresponding to positive- and negative-energy branches (see also Fig. 1.8). These play the role of the conduction and valence bands of the surface carriers. The conical energy spectrum (1.22) is reminiscent of that of an ultra-relativistic Dirac electron. For this reason, the surface states in 3DTIs are frequently called Dirac fermions.

The family of materials and heterostructures that can host surface states is pretty large. The surface states were first predicted at interfaces between normal and inverted bulk semiconductors (see, e.g., Refs. [20, 21]). More recently, a number of other

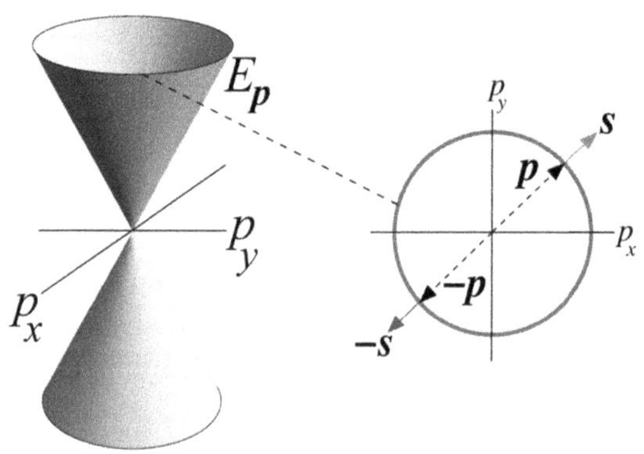

Figure 1.8 Three-dimensional plot of surface state spectrum (1.22) (left) and an example of the pair of protected states on a constant-energy line (right). Arrows illustrate the locking of momentum $\boldsymbol{p}$ and spin $\boldsymbol{s}$ vectors.

candidate materials to host a three-dimensional TI phase have been identified, for example, the semiconducting alloy $Bi_{1-x}Sb_x$ [9], strained layers of α-Sn and HgTe [9], tetradymite compounds, for example, Bi_2Se_3, Bi_2Te_3 and Sb_2Te_3 [22], thallium-based ternary chalcogenides $TlBiTe_2$ and $TlBiSe_2$ [23–25] as well as Pb-based layered chalcogenides [26, 27]. Experimentally, topological surface states have been observed by means of angle-resolved photoemission spectroscopy (ARPES) in $Bi_{1-x}Sb_x$ [11], Bi_2Se_3 [28], Bi_2Te_3 [29], $TlBiSe_2$ [30–32], $TlBiTe_2$ [32], strained HgTe [33], $Pb(Bi_{1-x}Sb_x)_2Te_4$ [34] and $PbBi_2Te_4$ [35].

It is important to keep in mind that the band structure of realistic TI materials is more complex than the Dirac cone described by Eq. (1.22). Away from the Dirac point $\boldsymbol{p} = 0$, the linear surface dispersion is modified by higher-order terms $\propto \boldsymbol{p}^2$. These break the *particle-hole symmetry*, that is, the symmetry between the positive- and negative-energy branches of Eq. (1.22), as illustrated in Fig. 1.9(left). In the tetradymite compounds Bi_2Se_3, Bi_2Te_3 and Sb_2Te_3, the surface bands become also anisotropic in momentum space due to *hexagonal warping* effects [36–39] [see also Fig. 1.9(middle)]. Furthermore, in thin films of TI materials the band topology undergoes a qualitative change: instead of a gapless Dirac cone, a

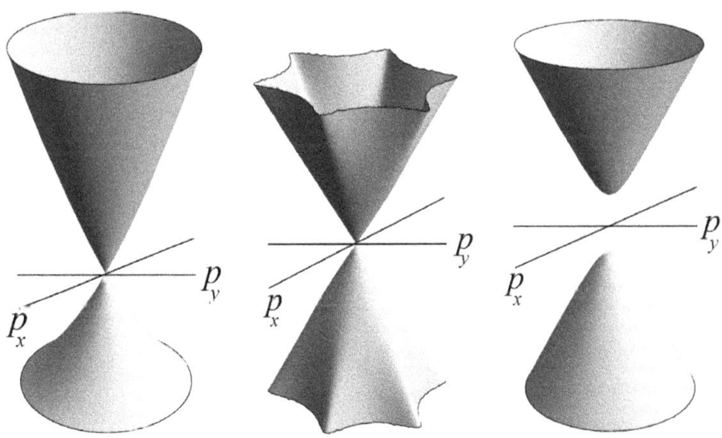

Figure 1.9 Illustration of topological states with broken particle-hole symmetry (left), hexagonal warping (middle) and energy gap (right).

semiconductor-type spectrum with an *energy gap* at $\boldsymbol{p} = 0$ is formed [see, e.g., Ref. [40, 41] and Fig. 1.9(right)]. Later, we shall discuss how the modifications of the band structure affect transport properties of the TIs. However, despite these modifications the identifying signature of the 3DTIs is still their Dirac-like dispersion (1.20) because it exhibits protected pair states $\boldsymbol{p}, \boldsymbol{s}$ and $-\boldsymbol{p}, -\boldsymbol{s}$, related to each other by the TR (see Fig. 1.8).

Chapter 2

Two-Dimensional Topological Insulators

In the preceding chapter, we used simple intuitive arguments to introduce the two-dimensional topological insulators (2DTIs) as the 2D systems with an insulating interior and conducting helical edge states (see Section 1.3). Now we intend to construct a theoretical model that would elucidate the topological origin of the helical edge states. We have already mentioned the original theoretical works on the 2DTIs in the context of graphene [1, 2] and semiconductor quantum wells [3, 4]. These models are closely related to the notion of Chern insulators introduced by Haldane [42]. Chern insulators are nontrivial band insulators in which broken time-reversal symmetry (TRS) induces a quantum Hall effect without any external magnetic field. This topological state is characterized by a gapless edge mode and the nonzero Thouless-Kohmoto-Nightingale-den Nijs (TKNN) invariant. These ingredients are instrumental for understanding the TR-invariant TIs as well. The chapter opens with a simple model of the Chern insulator and the derivation of the TKNN topological invariant. This will lead us to a TR-invariant theory of 2DTIs based on a low-energy continuum model for the 2D Dirac matter. We shall see that the helical edge states emerge from a local boundary condition, that is topologically equivalent to a band-gap domain wall describing the inversion of the band structure at the edge.

Topological Insulators: The Physics of Spin Helicity in Quantum Transport
Grigory Tkachov

ISBN 978-981-4613-25-5 (Hardcover), 978-981-4613-26-2 (eBook)
www.panstanford.com

2.1 Chern Insulator

2.1.1 *Edge Mode and Band Inversion*

We consider a 2D Dirac system described by the Hamiltonian

$$\hat{h} = \boldsymbol{\sigma} \cdot \boldsymbol{d}, \qquad \boldsymbol{d} = (\hbar v k_x, \hbar v k_y, m). \tag{2.1}$$

It is the Hamiltonian of Eq. (1.20) with an extra term, $m\sigma_z$, introducing a finite Dirac mass (taken, for concreteness, positive throughout). The eigenstates and -values of the Hamiltonian can be found from the equation

$$E\psi_{\boldsymbol{k}} = \boldsymbol{\sigma} \cdot \boldsymbol{d}\, \psi_{\boldsymbol{k}} = \sqrt{A^2\boldsymbol{k}^2 + m^2}\, (\boldsymbol{\sigma} \cdot \boldsymbol{u})\psi_{\boldsymbol{k}}, \tag{2.2}$$

where we defined a unit vector $\boldsymbol{u} = \boldsymbol{d}/|\boldsymbol{d}| = (u_\parallel \cos\phi, u_\parallel \sin\phi, u_z)$ in the polar coordinates $k_x = k\cos\phi$ and $k_y = k\sin\phi$ and with

$$u_\parallel = \sqrt{1 - u_z^2}, \qquad u_z = \frac{m}{\sqrt{(\hbar v \boldsymbol{k})^2 + m^2}}. \tag{2.3}$$

We can choose the solutions to be the eigenstates of the spin matrix $\boldsymbol{\sigma} \cdot \boldsymbol{u}$. It is the projection of the spin on the direction of the unit vector $\boldsymbol{u}$. This matrix is called helicity, and its eigenstates satisfy the equation

$$(\boldsymbol{\sigma} \cdot \boldsymbol{u})\psi_{\boldsymbol{k}s} = s\psi_{\boldsymbol{k}s}, \tag{2.4}$$

with real eigenvalues $s = \pm 1$. The solutions of Eq. (2.4), normalized to unity, are

$$\psi_{\boldsymbol{k}s} = \frac{\sigma_0 + s\boldsymbol{\sigma} \cdot \boldsymbol{u}}{\sqrt{2(1 + su_z)}} \begin{bmatrix} 1 \\ 0 \end{bmatrix} = \frac{1}{\sqrt{2}} \begin{bmatrix} \sqrt{1 + su_z} \\ se^{i\phi}\sqrt{1 - su_z} \end{bmatrix}. \tag{2.5}$$

The corresponding eigenvalues follow from Eqs. (2.2) and (2.4),

$$E_{\boldsymbol{k}s} = s\sqrt{A^2\boldsymbol{k}^2 + m^2}. \tag{2.6}$$

The positive ($s = +1$)- and negative ($s = -1$)- energy states correspond to the bulk conduction and valence bands, respectively. The mass term opens an energy gap of $2m$ at the Dirac point $\boldsymbol{k} = 0$ between the bands (see also Fig. 2.1a). Therefore, in the bulk the system behaves as a 2D band insulator. This is a nontrivial insulator, though, as it appears to have a gapless mode localized at the edge of the system. Such localized zero modes were first found in relativistic

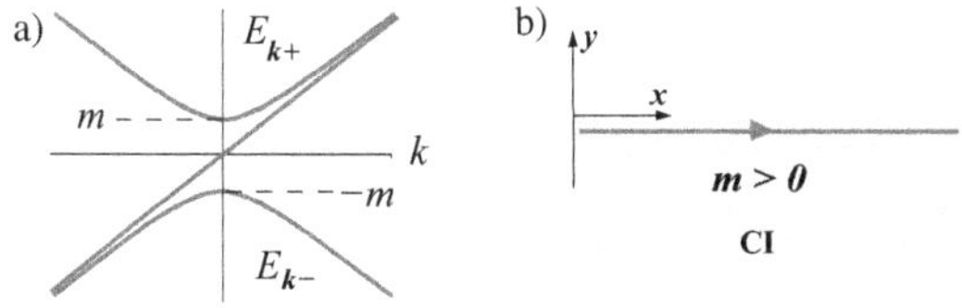

Figure 2.1 (a) Spectrum of the Chern insulator (CI) [see Eqs. (2.6) and (2.18)] and (b) system geometry.

theory of 1D fermions [43]. To see how this comes about in our system, we consider the real-space Dirac equation

$$[-i\hbar v(\sigma_x \partial_x + \sigma_y \partial_y) + m\sigma_z]\psi(\boldsymbol{r}) = E\,\psi(\boldsymbol{r}) \tag{2.7}$$

in a 2D semi-space chosen as $-\infty < y \leq 0$ and $-\infty < x < \infty$ [see also Fig. 2.1b]. The spinor $\psi(\boldsymbol{r})$ must satisfy a boundary condition that ensures vanishing of the normal component of the particle current density, $j_y(\boldsymbol{r})$, at the edge $y = 0$,

$$j_y(x, y = 0). \tag{2.8}$$

In order to derive the appropriate boundary condition, we recall the expression for the particle current density, $\boldsymbol{j}(\boldsymbol{r})$, in terms of an expectation value of the velocity operator, $\hat{\boldsymbol{v}}$:

$$\boldsymbol{j}(\boldsymbol{r}) = \tilde{\psi}(\boldsymbol{r})\hat{\boldsymbol{v}}\psi(\boldsymbol{r}), \qquad \hat{\boldsymbol{v}} = (\boldsymbol{r}\hat{h} - \hat{h}\boldsymbol{r})/i\hbar, \tag{2.9}$$

where $\tilde{\psi}(\boldsymbol{r})$ denotes the conjugated spinor,

$$\tilde{\psi}(\boldsymbol{r}) = [\psi^*_\uparrow(\boldsymbol{r}), \psi^*_\downarrow(\boldsymbol{r})]. \tag{2.10}$$

From Eqs. (2.1) and (2.9) we find

$$j_y(\boldsymbol{r}) = v\tilde{\psi}(\boldsymbol{r})\sigma_y\psi(\boldsymbol{r}) = iv[\psi^*_\downarrow(\boldsymbol{r})\psi_\uparrow(\boldsymbol{r}) - \psi^*_\uparrow(\boldsymbol{r})\psi_\downarrow(\boldsymbol{r})]. \tag{2.11}$$

A suitable boundary condition is

$$\psi_\uparrow(x, y = 0) = \psi_\downarrow(x, y = 0), \tag{2.12}$$

which in a spinor form reads

$$\psi(x, y = 0) = \sigma_x\psi(x, y = 0). \tag{2.13}$$

We seek the solution in the form of a plane wave propagating along the edge with a wave vector k and decaying exponentially in the bulk on a length λ:

$$\psi(x, y) = \Phi\, e^{ikx + y/\lambda}, \tag{2.14}$$

$$\Phi = \sigma_x\Phi. \tag{2.15}$$

To satisfy the boundary condition (2.13) the position-independent spinor Φ is subject to constraint (2.15). Inserting Eq. (2.14) into Eq. (2.7) we have

$$\hbar v k\Phi + (m - \hbar v/\lambda)\,\sigma_z\Phi = E\,\Phi, \tag{2.16}$$

where we used Eq. (2.15) and $\sigma_y\Phi = -i\sigma_z\Phi$. Setting the expression in the brackets to zero, we find the decay length

$$\lambda = \hbar v/m. \tag{2.17}$$

Then, Eq. (2.16) yields the edge-state spectrum

$$E_k = \hbar v k. \tag{2.18}$$

The obtained solution has several notable properties:

- It is gapless, that is, it spans the energy range within the gap, $-m < E < m$, which is forbidden for the bulk states (see Fig. 2.1a). The fact that we obtained a single chiral mode reflects broken TRS. The TRS invariance is violated by the mass term $m\sigma_z$ which does not commute with the TR operator $\mathbb{T} = i\sigma_y C$.
- The occurrence of a gapless mode is generally associated with the inversion of the band structure at the edge [43]. Indeed, any state $\psi(x, y)$ in the negative semispace $y < 0$ can be extended to the positive semispace $y > 0$ by the parity transformation

$$\psi'(x, y) = \sigma_x\psi(x, -y). \tag{2.19}$$

The boundary condition (2.13) is invariant under this operation, whereas in Dirac equation (2.7) the mass term changes its sign

$$[-i\hbar v(\sigma_x\partial_x + \sigma_y\partial_y) - m\sigma_z]\psi'(x, y) = E\,\psi'(x, y). \tag{2.20}$$

Such a mirror system has the same conduction- and valence-band spectra as in Eq. (2.6), but the ordering of the conduction and valence bands is opposite to the original (see Fig. 2.2). In other words, there is a band inversion at the edge accompanied by the gap closing and, hence, the appearance of a gapless mode. Our model is equivalent to a step-like band inversion at the edge. Similar results can be obtained for a smoothly varying $m(y)$ [43]. From the parity relation (2.19), we also see that the edge mode must be an eigenstate

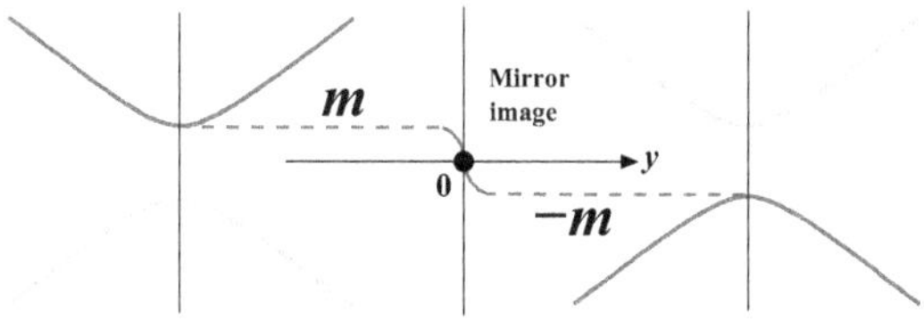

Figure 2.2 Illustration of the band inversion and gapless mode at the CI edge. The edge is equivalent to a boundary between a domain with mass m and its mirror image with $-m$ obtained by parity transformation $y \to -y$ (see also text).

of the spin projection σ_x since at $y = 0$ the functions $\psi'(x, 0)$ and $\psi(x, 0)$ describe the same state and, therefore, must coincide, which yields the boundary condition (2.13).

Most important, a bulk band insulator with a propagating gapless mode at the edge is topologically distinct from an ordinary band insulator. As we shall see, the two states of the electronic matter are distinguished by a certain topological quantum number that vanishes for an ordinary band insulator and is nonzero for the nontrivial one. Before going to that point, we need to introduce the notions of the Berry phase and Berry curvature for our system.

2.1.2 *Berry's Phase and Curvature*

Berry's phase is a quantum phase factor that appears in the course of an adiabatic evolution of a quantum state [44]. Let us consider an evolution of a bulk state $|\psi_{\boldsymbol{k}s}\rangle$ (2.5) under an adiabatic rotation of the wave vector, which we model by a time-dependent function $\boldsymbol{k}(t)$ as sketched in Fig. 2.3. It is assumed that at any time t the evolving state $|\chi(t)\rangle$ remains an eigenstate of the Hamiltonian $\hat{h}(\boldsymbol{k}(t))$ that depends on time only through the function $\boldsymbol{k}(t)$. Then, the equation of motion for $|\chi(t)\rangle$ reduces to

$$i\hbar\partial_t|\chi(t)\rangle = E_{\boldsymbol{k}(t)s}\,|\chi(t)\rangle. \tag{2.21}$$

Its solution must be proportional to $|\psi_{\boldsymbol{k}(t)s}\rangle$, so that we use the ansatz

$$|\chi(t)\rangle = e^{-i\theta(t)}|\psi_{\boldsymbol{k}(t)s}\rangle, \qquad \langle\psi_{\boldsymbol{k}(t)s}|\psi_{\boldsymbol{k}(t)s}\rangle = 1, \tag{2.22}$$

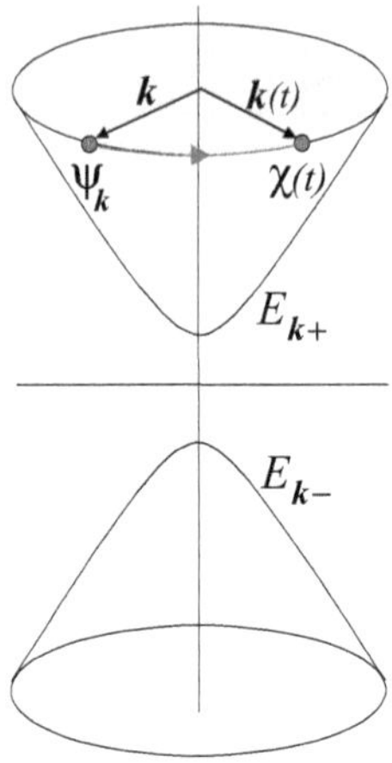

Figure 2.3 Illustration of an evolution of a bulk state in a CI under an adiabatic rotation of the wave vector modeled by a time-dependent function $\boldsymbol{k}(t)$ (see also text).

where $\theta(t)$ is the unknown phase acquired by time t. Inserting Eq. (2.22) into Eq. (2.21) yields

$$i\hbar\partial_t|\psi_{\boldsymbol{k}(t)s}\rangle + \hbar\dot{\theta}(t)|\psi_{\boldsymbol{k}(t)s}\rangle = E_{\boldsymbol{k}(t)s}|\psi_{\boldsymbol{k}(t)s}\rangle. \tag{2.23}$$

Applying $\langle\psi_{\boldsymbol{k}(t)s}|$ from the left, we have

$$\dot{\theta}(t) = \frac{1}{\hbar}E_{\boldsymbol{k}(t)s} - i\langle\psi_{\boldsymbol{k}(t)s}|\partial_t|\psi_{\boldsymbol{k}(t)s}\rangle, \tag{2.24}$$

and, finally, the solution for the phase

$$\theta(t) = \frac{1}{\hbar}\int_0^t E_{\boldsymbol{k}(t')s}dt' - i\int_0^t \langle\psi_{\boldsymbol{k}(t')s}|\partial_{t'}|\psi_{\boldsymbol{k}(t')s}\rangle dt'. \tag{2.25}$$

Here the first term is the ordinary dynamic phase, whereas the second one originates from the adiabatic evolution of $|\psi_{\boldsymbol{k}(t)s}\rangle$. Upon completing the full cycle at time t_{cycle}, the second contribution becomes

$$\gamma_s = -i\int_0^{t_{cycle}} \langle\psi_{\boldsymbol{k}(t')s}|\partial_{t'}|\psi_{\boldsymbol{k}(t')s}\rangle dt' = -i\oint\langle\psi_{\boldsymbol{k}s}|\nabla_{\boldsymbol{k}}|\psi_{\boldsymbol{k}s}\rangle\cdot d\boldsymbol{k}, \tag{2.26}$$

where the time integration is converted into a path integral around $\boldsymbol{k} = 0$ in momentum space. This geometrical phase is the analogue of Berry's phase for our system. Defining the geometrical vector potential (Berry connection) by

$$\boldsymbol{A}_{\boldsymbol{k}s} = -i\langle\psi_{\boldsymbol{k}s}|\nabla_{\boldsymbol{k}}|\psi_{\boldsymbol{k}s}\rangle, \tag{2.27}$$

one can express the Berry phase in the form analogous to the magnetic flux,

$$\gamma_s = \oint \boldsymbol{A}_{\boldsymbol{k}s}\cdot d\boldsymbol{k} = \int (\nabla_{\boldsymbol{k}}\times\boldsymbol{A}_{\boldsymbol{k}s})_z\, dk_x dk_y, \tag{2.28}$$

where the Berry curvature $(\nabla_{\boldsymbol{k}}\times\boldsymbol{A}_{\boldsymbol{k}s})_z$ plays the role of the magnetic field perpendicular to the area swept out by the integration path.

Next, we take advantage of the axial symmetry of the problem and write the Berry connection in the polar coordinate system

$$\boldsymbol{A}_{\boldsymbol{k}s} = A^{(k)}_{\boldsymbol{k}s}\hat{\boldsymbol{k}} + A^{(\phi)}_{\boldsymbol{k}s}\hat{\boldsymbol{\phi}}, \tag{2.29}$$

where $\hat{\boldsymbol{k}}$ and $\hat{\boldsymbol{\phi}}$ are the radial and azimuthal basis unit vectors, respectively. With the wave functions given by Eq. (2.5), the radial and azimuthal projections of $\boldsymbol{A}_{\boldsymbol{k}s}$ are

$$A^{(k)}_{\boldsymbol{k}s} = -i\langle\psi_{\boldsymbol{k}s}|\frac{\partial}{\partial k}|\psi_{\boldsymbol{k}s}\rangle = 0, \tag{2.30}$$

$$A^{(\phi)}_{\boldsymbol{k}s} = -i\langle\psi_{\boldsymbol{k}s}|\frac{\partial}{k\partial\phi}|\psi_{\boldsymbol{k}s}\rangle = \frac{1-su_z}{2k}. \tag{2.31}$$

By symmetry, only the azimuthal projection is nonzero, which yields the Berry phase

$$\gamma_s = \oint \boldsymbol{A}_{\boldsymbol{k}s}\cdot d\boldsymbol{k} = 2\pi k A^{(\phi)}_{\boldsymbol{k}s} = \pi(1-su_z) \tag{2.32}$$

$$= \pi\left(1 - \frac{sm}{\sqrt{\hbar^2v^2\boldsymbol{k}^2+m^2}}\right), \tag{2.33}$$

and the Berry curvature

$$(\nabla_{\boldsymbol{k}}\times\boldsymbol{A}_{\boldsymbol{k}s})_z = -\frac{s}{2k}\frac{\partial u_z}{\partial k} = \frac{s(\hbar v)^2}{2}\frac{m}{(\hbar^2v^2\boldsymbol{k}^2+m^2)^{3/2}} \tag{2.34}$$

$$= (\hbar v)^2 \times \frac{s}{2}\frac{\boldsymbol{d}\cdot\boldsymbol{z}}{d^3}. \tag{2.35}$$

We note that for a gapless Dirac cone the Berry phase is the universal number of π. The nonuniversal contribution is proportional to the mass term m and has opposite signs for the conduction and valence bands. Interestingly, the Berry curvature (2.35) is analogous to the field of a monopole positioned at the point $\boldsymbol{k}=0$ with the fractional strength $s/2$ in units of $(\hbar v)^2$. This observation is intimately related to the quantum Hall effect and topological invariants which are discussed below.

2.1.3 *Thouless-Kohmoto-Nightingale-den Nijs formula for the Hall Conductivity. Topological Invariants*

Our next goal is to introduce the notion of a topological invariant. It allows for a rigorous classification of topological phases of an electronic system and is insensitive to continuous deformations of the electronic band structure. For systems with broken TRS, the topological invariant is directly related to an observable, the quantized Hall conductivity. In order to illustrate all these points we shall derive an important formula, first proposed by Thouless, Kohmoto, Nightingale, and den Nijs (TKNN) [45], that expresses the quantum Hall conductivity in terms of the Berry curvature of all occupied electronic bands of the system.

To calculate the electric conductivity, we employ the linear response theory with respect to an external uniform electric field $\mathcal{E}e^{-i\omega t}$ with the frequency ω. The details of this standard approach are well described in literature (Refs. [46, 47]). The conductivity is a 2×2 tensor whose components are given by the Kubo formula

$$\sigma_{\alpha\beta}(\omega) = -\frac{ie^2\hbar}{a}\sum_{l,m}\frac{f(E_l) - f(E_m)}{E_l - E_m}\frac{\langle l|v_\alpha|m\rangle\langle m|v_\beta|l\rangle}{\hbar\omega + E_l - E_m}, \tag{2.36}$$

where $\langle l|v_\alpha|m\rangle$ is the matrix element of the α-projection of the velocity operator

$$v_\alpha = (x_\alpha\hat{h} - \hat{h}x_\alpha)/i\hbar \tag{2.37}$$

between quantum states $|m\rangle$ and $|l\rangle$ with energies E_m and E_l:

$$\hat{h}|m, l\rangle = E_{m,l}|m, l\rangle. \tag{2.38}$$

The indices m and l include the full sets of quantum numbers, while α and β refer to the spatial coordinates x and y; $f(E)$ is the Fermi–Dirac distribution function, e is the electron charge, and a is the area of the system.

We are interested in the transverse (Hall) conductivity σ_{xy} in the limit of zero frequency and temperature

$$\sigma_{xy} = -\frac{ie^2\hbar}{a}\sum_{l,m}[\Theta(\mu - E_l) - \Theta(\mu - E_m)]\frac{\langle l|v_x|m\rangle\langle m|v_y|l\rangle}{(E_l - E_m)^2}, \quad (2.39)$$

and with the Fermi level, μ, lying between two arbitrary energy levels E_m and E_l [$\Theta(E)$ is the unit-step function]. Two level configurations are possible:

$$E_l < \mu < E_m, \quad (2.40)$$

and

$$E_m < \mu < E_l. \quad (2.41)$$

For the former the difference of the Θ functions in Eq. (2.39) is $+1$, whereas for the latter this difference is -1. Summing up the two possibilities, we can write

$$\sigma_{xy} = -\frac{ie^2\hbar}{a}\sum_{E_l<\mu,\, E_m>\mu}\frac{\langle l|v_x|m\rangle\langle m|v_y|l\rangle - \langle l|v_y|m\rangle\langle m|v_x|l\rangle}{(E_l - E_m)^2}. \quad (2.42)$$

The next step is to transform the velocity matrix elements with the help of the following identities

$$\langle m|v_\alpha|l\rangle = \frac{\langle m|x_\alpha\,\hat{h}|l\rangle - \langle m|\hat{h}\,x_\alpha|l\rangle}{i\hbar} = \frac{E_l - E_m}{i\hbar}\langle m|x_\alpha|l\rangle \quad (2.43)$$

$$= \frac{E_l - E_m}{i\hbar}\langle m|i\,\partial_{k_\alpha}|l\rangle = \frac{E_l - E_m}{\hbar}\langle m|\partial_{k_\alpha}l\rangle, \quad (2.44)$$

and

$$\langle l|v_\alpha|m\rangle = \frac{\langle l|x_\alpha\,\hat{h}|m\rangle - \langle l|\hat{h}\,x_\alpha|m\rangle}{i\hbar} = \frac{E_m - E_l}{i\hbar}\langle l|x_\alpha|m\rangle \quad (2.45)$$

$$= \frac{E_m - E_l}{i\hbar}\langle l|i\,\partial_{k_\alpha}|m\rangle = -\frac{E_m - E_l}{\hbar}\langle\partial_{k_\alpha}l|m\rangle. \quad (2.46)$$

Here, we used Eq. (2.38) and its conjugate $\langle m, l|\hat{h} = E_{m,l}\langle m, l|$ and expressed the coordinate operator in the wave-vector space as $x_\alpha = i\,\partial_{k_\alpha}$. In Eq. (2.44) $|\partial_{k_\alpha}l\rangle$ denotes the derivative of the ket $|l\rangle$ with respect to k_α, while in Eq. (2.46) the derivative is transferred to the bra $\langle l|$ by means of $\langle l|\partial_{k_\alpha}m\rangle = -\langle\partial_{k_\alpha}l|m\rangle$. The latter step is enabled by the hermiticity of the coordinate operator. It is worth reminding that for the coordinate operator to be hermitian in $\boldsymbol{k}$ space the entire Brillouin zone of the crystal must be taken into account with the periodic boundary conditions at the zone edges. For simplicity, we

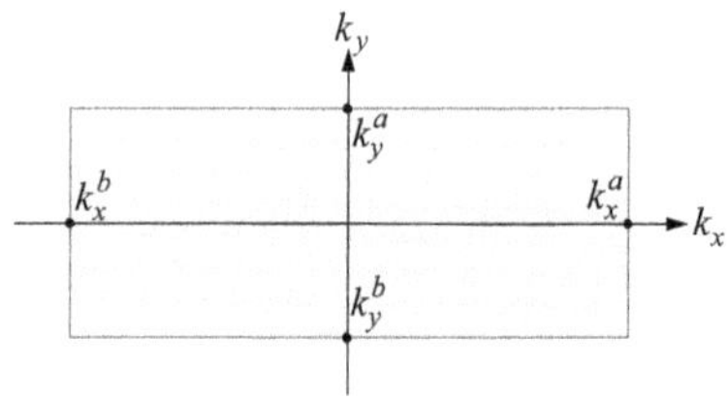

Figure 2.4 Schematic of the Brillouin zone assumed in the derivation of the TKNN formula. The electronic wave function is periodic in both k_x- and k_y-directions [see also Eq. (2.47)].

assume a rectangular Brillouin zone with the following periodic boundary conditions (see also Fig. 2.4)

$$|m\rangle_{k^a_{x,y}} = |m\rangle_{k^b_{x,y}}. \tag{2.47}$$

Equations (2.44) and (2.46) allow us to express the Hall conductivity in terms of the eigenfunctions only

$$\sigma_{xy} = -\frac{ie^2}{\hbar a}\sum_{\boldsymbol{k}}\sum_{E_l<\mu, E_m>\mu}\left[\langle\partial_{k_x} l|m\rangle\langle m|\partial_{k_y} l\rangle - \langle\partial_{k_y} l|m\rangle\langle m|\partial_{k_x} l\rangle\right]. \tag{2.48}$$

Next, we eliminate the unoccupied states with $E_m > \mu$ using the completeness condition

$$\sum_{E_m>\mu}|m\rangle\langle m| = 1 - \sum_{E_m<\mu}|m\rangle\langle m|, \tag{2.49}$$

which yields

$$\sigma_{xy} = -\frac{ie^2}{\hbar a}\sum_{\boldsymbol{k}}\sum_{E_l<\mu}\left[\langle\partial_{k_x} l|\partial_{k_y} l\rangle - \langle\partial_{k_y} l|\partial_{k_x} l\rangle\right] \tag{2.50}$$

$$+\frac{ie^2}{\hbar a}\sum_{\boldsymbol{k}}\sum_{E_l<\mu, E_m<\mu}\left[\langle\partial_{k_x} l|m\rangle\langle m|\partial_{k_y} l\rangle - \langle\partial_{k_y} l|m\rangle\langle m|\partial_{k_x} l\rangle\right].$$

The second term vanishes due the periodicity of the boundary conditions (2.47). Indeed, using again $\langle\partial_{k_x} l|m\rangle = -\langle l|\partial_{k_x} m\rangle$ and $\langle m|\partial_{k_y} l\rangle = -\langle\partial_{k_y} m|l\rangle$ we can write

$$\langle\partial_{k_x} l|m\rangle\langle m|\partial_{k_y} l\rangle - \langle\partial_{k_y} l|m\rangle\langle m|\partial_{k_x} l\rangle$$
$$= \langle\partial_{k_y} m|l\rangle\langle l|\partial_{k_x} m\rangle - \langle\partial_{k_y} l|m\rangle\langle m|\partial_{k_x} l\rangle, \tag{2.51}$$

which is odd under the exchange of l and m and, hence, vanishes under the symmetric double summation over l and m. The result is

$$\sigma_{xy} = -\frac{ie^2}{\hbar a} \sum_{\boldsymbol{k}} \sum_{E_l<\mu} \left[\langle \partial_{k_x} l | \partial_{k_y} l \rangle - \langle \partial_{k_y} l | \partial_{k_x} l \rangle \right]. \tag{2.52}$$

Defining the Berry connection of the occupied bands as

$$A^{(x)}_{\boldsymbol{k}} = -i \sum_{E_l<\mu} \langle l | \partial_{k_x} | l \rangle, \qquad A^{(y)}_{\boldsymbol{k}} = -i \sum_{E_l<\mu} \langle l | \partial_{k_y} | l \rangle, \tag{2.53}$$

we can cast Eq. (2.52) into the form

$$\begin{aligned} \sigma_{xy} &= \frac{e^2}{h} \int \frac{dk_x dk_y}{2\pi} \left(\partial_{k_x} A^{(y)}_{\boldsymbol{k}} - \partial_{k_y} A^{(x)}_{\boldsymbol{k}} \right) \\ &= \frac{e^2}{h} \int \frac{dk_x dk_y}{2\pi} (\nabla_{\boldsymbol{k}} \times \boldsymbol{A}_{\boldsymbol{k}})_z, \end{aligned} \tag{2.54}$$

where the integration extends over the entire Brillouin zone of the system. This is the TKNN formula. It establishes a general link between the Hall conductivity and the topology of the electronic band structure. Two aspects of Eq. (2.54) merit special attention:

- The Hall conductivity is nonzero due to broken TRS and is linked to chiral gapless edge states. To illustrate this point, let us take the model of the CI introduced earlier in Section 2.1.1. Since the Fermi level must be in the band gap, there is one occupied band - the valence one. Therefore, the Berry connection in Eq. (2.53) is identified as $\boldsymbol{A}_{\boldsymbol{k}s}$ with $s = -1$ from Section 2.1.2. Using Eq. (2.34) for the Berry curvature, we can easily do the integral in Eq. (2.54)

$$\sigma_{xy} = \frac{e^2}{h} \int_0^\infty \frac{2\pi k dk}{2\pi} \frac{1}{2k} \frac{\partial u_z}{\partial k} = -\frac{e^2}{h} \frac{u_z(k=0)}{2} = -\frac{e^2}{h} \frac{\text{sgn}(m)}{2}. \tag{2.55}$$

  As we see, the Hall conductivity is induced by the TRS-breaking mass term. If we set $m = 0$ before evaluating the integral, the result is $\sigma_{xy} = 0$. Also, Eq. (2.55) relates σ_{xy} to the contribution of the midgap state with $k = 0$. The only state that exists in the gap is the edge mode (see Section 2.1.1). Therefore, the Hall conductivity can be associated with propagating edge states.

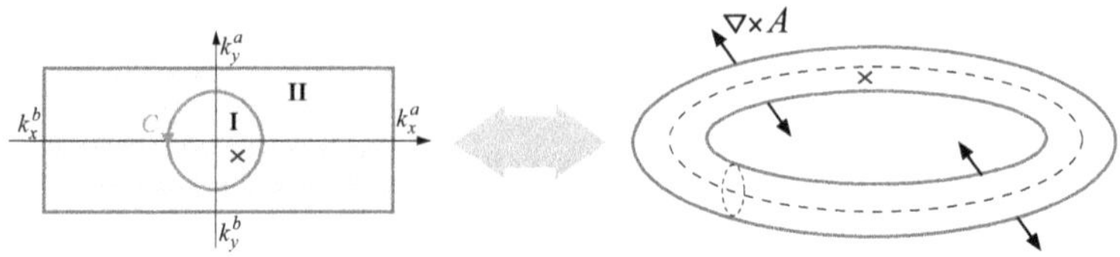

Figure 2.5 Mapping the Brillouin zone to the surface of a torus. The cross indicates a zero of the electronic wave function that results in the nontrivial winding of the phase (see also text).

- The Hall conductivity is quantized in integer units of the conductance quantum e^2/h. We cannot see this from the CI model for the reasons that will be clear later. Fortunately, the proof rests only on topological properties of the TKNN formula and is free of any model specifics [45, 48]. In view of the periodic boundary conditions (2.47) we can connect the edges of the Brillouin zone in both k_x- and k_y- directions, so that it maps to the surface of a torus and, as such, has no boundary (see also Fig. 2.5). This allows us to divide the Brillouin zone into two closed manifolds, denoted below I and II, and write Eq. (2.54) as

$$\sigma_{xy} = \frac{e^2}{h}\left[\int_I \frac{dk_x dk_y}{2\pi}(\nabla_{\boldsymbol{k}} \times \boldsymbol{A}_{\boldsymbol{k}})_z + \int_{II} \frac{dk_x dk_y}{2\pi}(\nabla_{\boldsymbol{k}} \times \boldsymbol{A}_{\boldsymbol{k}})_z\right]. \tag{2.56}$$

The area integrals can be converted into the line ones along the the common boundary C, but with the opposite signs because the circulation direction of C is opposite for manifolds I and II:

$$\sigma_{xy} = \frac{e^2}{h}\frac{1}{2\pi}\left[\oint_C \boldsymbol{A}^I_{\boldsymbol{k}} \cdot d\boldsymbol{k} - \oint_C \boldsymbol{A}^{II}_{\boldsymbol{k}} \cdot d\boldsymbol{k}\right]. \tag{2.57}$$

The gauge invariance allows the Berry connections $\boldsymbol{A}^I_{\boldsymbol{k}}$ and $\boldsymbol{A}^{II}_{\boldsymbol{k}}$ to differ only by a gradient of an arbitrary phase of the wave function,

$$\boldsymbol{A}^I_{\boldsymbol{k}} - \boldsymbol{A}^{II}_{\boldsymbol{k}} = \nabla_{\boldsymbol{k}}\theta, \tag{2.58}$$

so that the Hall conductivity is proportional to the winding number of the phase of the electronic wave function, n:

$$\sigma_{xy} = \frac{e^2}{h}n, \qquad n = \frac{1}{2\pi}\oint_C \nabla_{\boldsymbol{k}}\theta \cdot d\boldsymbol{k}. \tag{2.59}$$

For a uniquely defined wave function, n is an integer, and hence, the Hall conductivity is quantized in integer units of e^2/h. This result has remarkable parallels with the quantization of the superfluid velocity and the magnetic flux in superconductors. In those systems, the nontrivial phase winding numbers $n \neq 0$ are associated with vortices–spatial singularities at which the condensate order parameter vanishes. In our case, the role of the vortices is played by the zeros of the electronic wave function for a given band in the wave-vector space. Thus, the quantized Hall conductivity is proportional to the total vorticity of the wave function in the Brillouin zone under broken TRS [45, 48].

The TKNN number n is determined solely by the electronic band structure:

$$n = \int \frac{dk_x dk_y}{2\pi} (\nabla_{\boldsymbol{k}} \times \boldsymbol{A}_{\boldsymbol{k}})_z \tag{2.60}$$

$$= -i \int \frac{dk_x dk_y}{2\pi} \sum_{E_l < \mu} \left[\langle \partial_{k_x} l | \partial_{k_y} l \rangle - \langle \partial_{k_y} l | \partial_{k_x} l \rangle \right]. \tag{2.61}$$

It is a topological invariant since it is robust against continuous deformations of the band structure that leave the number of the occupied bands and, hence, the number of the zeros in the Brillouin zone unchanged. In particular, Eq. (2.61) shows that the transition from the $n = 0$ to $n \neq 0$ state is a topological one since the topology of the Brillouin zone changes from singly to multiply connected. There is also a deep parallel with differential geometry in which n in Eq. (2.60) is identified as the first Chern number of a principal $U(1)$ bundle over a torus.

To conclude this section, we need to clarify why our CI model predicts a half-integer Hall conductivity [see Eq. (2.55)] instead of the integer one (2.59). The reason for this discrepancy is that the two-component Dirac model introduced in Section 2.1.1 is insufficient to represent a crystalline material. According to the general theorem proved by Nielsen and Ninomiya [49], Dirac fermions arranged on a crystalline lattice come only in pairs, each having a two-component spinor function. In other words, in order to represent all important states of the Brillouin zone the fermion content of our CI model must be at least twice as large, which would yield the correct integer-quantized Hall conductivity. This has been explicitly shown by Haldane in his model of the quantum Hall effect on a 2D honeycomb lattice [42]. Nonetheless, the CI model used so far is instrumental for understanding the TR invariant TIs considered in the rest of the book.

2.2 Time-Reversal Invariant Topological Insulators

2.2.1 *2D Dirac Matter. Effective Hamiltonian and Boundary Conditions*

Similar to the CI, the 2DTIs must have a band gap in the bulk, but the gap opening should occur without breaking the TRS. The model satisfying these requirements can be introduced by rather simple reasoning. Let us consider two 2D electronic systems at the top and bottom surfaces of a 3D slab with an insulating interior [see Fig. 2.6(left)]. Each system is described by the gapless TR-invariant Dirac Hamiltonian (1.20). As the thickness of the slab is reduced, the two systems become coupled by tunneling through the insulating layer between them. In the quasi-2D limit, when the top and bottom states overlap [see Fig. 2.6(right)], the whole system can be described by an equation

$$\hat{H}\,\psi(\boldsymbol{r}) = E\,\psi(\boldsymbol{r}), \qquad \boldsymbol{r} = (x, y, 0), \tag{2.62}$$

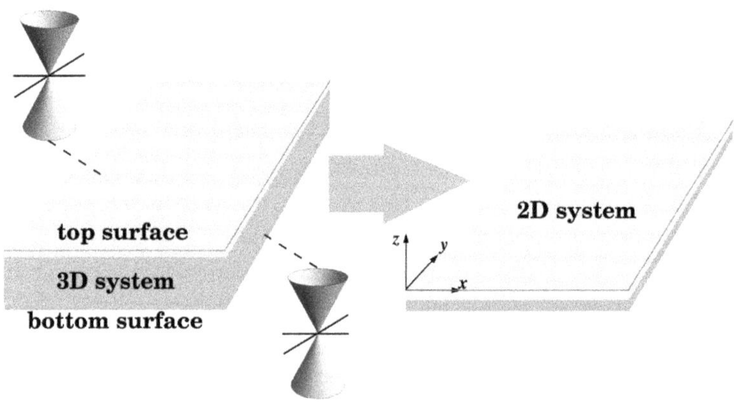

Figure 2.6 Schematic of a 3D insulator with top and bottom surface Dirac systems (left) and of its quasi-2D counterpart (right).

for a four-component wave function, $\psi(\boldsymbol{r})$, in the basis of the original top (t) and bottom (b) states each having two spin components $\uparrow$ and $\downarrow$:

$$\psi(\boldsymbol{r}) = \begin{bmatrix} \psi_{t\uparrow}(\boldsymbol{r}) \\ \psi_{t\downarrow}(\boldsymbol{r}) \\ \psi_{b\uparrow}(\boldsymbol{r}) \\ \psi_{b\downarrow}(\boldsymbol{r}) \end{bmatrix}. \tag{2.63}$$

The Hamiltonian $\hat{H}$ in Eq. (2.62) is a 4×4 matrix given by

$$\hat{H} = \begin{bmatrix} -i\hbar v(\sigma_x \partial_x + \sigma_y \partial_y) & \Delta\sigma_0 \\ \Delta\sigma_0 & +i\hbar v(\sigma_x \partial_x + \sigma_y \partial_y) \end{bmatrix}. \tag{2.64}$$

Here, the diagonal entries are the real-space Hamiltonians of the originally decoupled t and b subsystems. The choice of the signs is dictated by the inversion symmetry of the slab. The coupling between the two subsystems is described by the off-diagonal entry $\Delta\sigma_0$, where σ_0 is a unit matrix in spin space, which corresponds to spin-independent tunneling. For a real parameter Δ, the whole system is manifestly TR invariant since its Hamiltonian does not change under the $\mathbb{T}$ operation

$$\mathbb{T}\hat{H}\mathbb{T}^{-1} = \hat{H}, \quad \mathbb{T} = \begin{bmatrix} i\sigma_y C & 0 \\ 0 & i\sigma_y C \end{bmatrix}. \tag{2.65}$$

At the same time, as we shall see later, the coupling Δ induces an energy gap in the electron spectrum. Thus, Hamiltonian (2.64) is a good starting point to search for the TR-invariant topological states. It is also worth noting that this Hamiltonian can be derived from the band structure theory of concrete TI materials, for example, HgTe/CdTe quantum wells [4] and thin films of Bi_2Se_3 [50, 51].

We are interested in the solutions that decay exponentially from the edge of the system, $y = 0$, into its interior in the 2D semi-space $-\infty < x < \infty$ and $0 \leq y < \infty$ [see also Fig. 2.6(left)]. The spinor $\psi(\boldsymbol{r})$ must satisfy a boundary condition that ensures the vanishing of the normal component of the particle current density, $j_y(\boldsymbol{r})$, at the edge $y = 0$. In order to derive such a boundary condition, we start with the expression for the particle current density, $\boldsymbol{j}(\boldsymbol{r})$, in terms of an expectation value of the velocity operator, $\hat{\boldsymbol{v}}$:

$$\boldsymbol{j}(\boldsymbol{r}) = \tilde{\psi}(\boldsymbol{r})\hat{\boldsymbol{v}}\psi(\boldsymbol{r}), \qquad \hat{\boldsymbol{v}} = (\boldsymbol{r}\hat{H} - \hat{H}\boldsymbol{r})/i\hbar, \tag{2.66}$$

where $\tilde{\psi}(\boldsymbol{r})$ denotes the conjugated spinor,

$$\tilde{\psi}(\boldsymbol{r}) = [\psi^*_{t\uparrow}(\boldsymbol{r}), \psi^*_{t\downarrow}(\boldsymbol{r}), \psi^*_{b\uparrow}(\boldsymbol{r}), \psi^*_{b\downarrow}(\boldsymbol{r})]. \tag{2.67}$$

From Eqs. (2.64) and (2.9) we find

$$\begin{aligned} j_y(\boldsymbol{r}) &= v\tilde{\psi}(\boldsymbol{r}) \begin{bmatrix} \sigma_y & 0 \\ 0 & -\sigma_y \end{bmatrix} \psi(\boldsymbol{r}) \\ &= iv[\psi^*_{t\downarrow}(\boldsymbol{r})\psi_{t\uparrow}(\boldsymbol{r}) - \psi^*_{t\uparrow}(\boldsymbol{r})\psi_{t\downarrow}(\boldsymbol{r})] \\ &+ iv[\psi^*_{b\uparrow}(\boldsymbol{r})\psi_{b\downarrow}(\boldsymbol{r}) - \psi^*_{b\downarrow}(\boldsymbol{r})\psi_{b\uparrow}(\boldsymbol{r})]. \end{aligned} \tag{2.68}$$

To fulfill the requirement $j_y(x, y = 0) = 0$ we impose the following boundary conditions:

$$\psi_{b\uparrow}(x, y = 0) = \psi_{t\downarrow}(x, y = 0), \tag{2.69}$$

$$\psi_{b\downarrow}(x, y = 0) = -\psi_{t\uparrow}(x, y = 0), \tag{2.70}$$

Importantly, this is the only choice compatible with the TRS. In order to check this we write Eqs. (2.69)–(2.70) in a spinor form

$$\psi(x, y = 0) = \hat{M}\psi(x, y = 0), \qquad \hat{M} = \begin{bmatrix} 0 & -i\sigma_y \\ i\sigma_y & 0 \end{bmatrix}. \tag{2.71}$$

Since $i\sigma_y$ commutes with the TR operator $i\sigma_y C$, matrix $\hat{M}$ in Eq. (2.71) is invariant under operation (2.65),

$$\mathbb{T}\hat{M}\mathbb{T}^{-1} = \hat{M}, \tag{2.72}$$

that is, the transformed wave function $\mathbb{T}\psi(\boldsymbol{r})$ satisfies the same boundary condition as in Eq. (2.71). If $\hat{M}$ contained any of the other two Pauli matrices, σ_x or σ_z, it would not commute with $\mathbb{T}$.

2.2.2 *Normal and Inverted Band Structure*

Before approaching the problem of edge states, it is instructive to discuss the bulk band structure of an unconfined 2D Dirac matter. For that purpose, we are going to diagonalize the Hamiltonian, $\hat{H}_0$, (2.64) by an appropriate unitary transformation, $\hat{U}$, of Eq. (2.62),

$$\hat{H}'\,\psi'(\boldsymbol{r}) = E\,\psi'(\boldsymbol{r}), \qquad \boldsymbol{r} = (x, y, 0), \tag{2.73}$$

where the new spinor $\psi'(\boldsymbol{r})$ and Hamiltonian $\hat{H}'$ are

$$\psi'(\boldsymbol{r}) = \hat{U}\,\psi(\boldsymbol{r}), \tag{2.74}$$

$$\hat{H}' = \hat{U}\,\hat{H}\,U^{-1}. \tag{2.75}$$

Choosing the unitary matrix in the form

$$\hat{U} = \frac{1}{\sqrt{2}}\begin{bmatrix} \sigma_0 & \sigma_z \\ \sigma_z & -\sigma_0 \end{bmatrix}, \qquad \hat{U}^{-1} = \hat{U}, \tag{2.76}$$

we write Eqs. (2.74) and (2.75) explicitly as

$$\psi'(\boldsymbol{r}) = \frac{1}{\sqrt{2}}\begin{bmatrix} \psi_{t\uparrow}(\boldsymbol{r}) + \psi_{b\uparrow}(\boldsymbol{r}) \\ \psi_{t\downarrow}(\boldsymbol{r}) - \psi_{b\downarrow}(\boldsymbol{r}) \\ \psi_{t\uparrow}(\boldsymbol{r}) - \psi_{b\uparrow}(\boldsymbol{r}) \\ -\psi_{t\downarrow}(\boldsymbol{r}) - \psi_{b\downarrow}(\boldsymbol{r}) \end{bmatrix} \equiv \begin{bmatrix} \psi_{s\uparrow}(\boldsymbol{r}) \\ \psi_{a\downarrow}(\boldsymbol{r}) \\ \psi_{a\uparrow}(\boldsymbol{r}) \\ \psi_{s\downarrow}(\boldsymbol{r}) \end{bmatrix}, \tag{2.77}$$

$$\hat{H}' = \begin{bmatrix} -i\hbar v(\sigma_x\partial_x + \sigma_y\partial_y) + \Delta\sigma_z & 0 \\ 0 & +i\hbar v(\sigma_x\partial_x + \sigma_y\partial_y) - \Delta\sigma_z \end{bmatrix}. \tag{2.78}$$

The new basis in Eq. (2.77) has simple meaning. It consists of all symmetric (s) and anti-symmetric (a) combinations of the original t and b states for the same spin projection. This basis has the advantage that the new Hamiltonian (2.78) is block diagonal. The coupling Δ enters as the mass term.

Ignoring, for a moment, the boundaries, we expand $\psi'(\boldsymbol{r})$ in plane waves in 2D wave-vector ($\boldsymbol{k}$) space,

$$\psi'(\boldsymbol{r}) = \sum_{\boldsymbol{k}} \psi'_{\boldsymbol{k}} e^{i\boldsymbol{kr}}, \tag{2.79}$$

where spinor $\psi'_{\boldsymbol{k}}$ satisfies the equation

$$\hat{H}'_{\boldsymbol{k}}\,\psi'_{\boldsymbol{k}} = E\,\psi'_{\boldsymbol{k}}, \tag{2.80}$$

with Hamiltonian $\hat{H}'_{\boldsymbol{k}}$ given by

$$\hat{H}'_{\boldsymbol{k}} = \begin{bmatrix} \hbar v\boldsymbol{\sigma}\cdot\boldsymbol{k} + \Delta\sigma_z & 0 \\ 0 & -\hbar v\boldsymbol{\sigma}\cdot\boldsymbol{k} - \Delta\sigma_z \end{bmatrix}. \tag{2.81}$$

For $\boldsymbol{k} = 0$ Eq. (2.80) assumes a simple form,

$$\Delta\sigma_z \begin{bmatrix} \psi_{s\uparrow\boldsymbol{k}=0} \\ \psi_{a\downarrow\boldsymbol{k}=0} \end{bmatrix} = E \begin{bmatrix} \psi_{s\uparrow\boldsymbol{k}=0} \\ \psi_{a\downarrow\boldsymbol{k}=0} \end{bmatrix}, \tag{2.82}$$

$$-\Delta\sigma_z \begin{bmatrix} \psi_{a\uparrow\boldsymbol{k}=0} \\ \psi_{s\downarrow\boldsymbol{k}=0} \end{bmatrix} = E \begin{bmatrix} \psi_{a\uparrow\boldsymbol{k}=0} \\ \psi_{s\downarrow\boldsymbol{k}=0} \end{bmatrix}. \tag{2.83}$$

Equations (2.82) and (2.83) each have two solutions being the eigenstates of σ_z. In total, there are four solutions:

$$\begin{bmatrix} \psi_{s\uparrow\boldsymbol{k}=0} \\ 0 \\ 0 \\ 0 \end{bmatrix} \quad \text{with} \quad E_{s\uparrow} = \Delta, \tag{2.84}$$

$$\begin{bmatrix} 0 \\ \psi_{a\downarrow\boldsymbol{k}=0} \\ 0 \\ 0 \end{bmatrix} \quad \text{with} \quad E_{a\downarrow} = -\Delta, \tag{2.85}$$

$$\begin{bmatrix} 0 \\ 0 \\ \psi_{a\uparrow\boldsymbol{k}=0} \\ 0 \end{bmatrix} \quad \text{with} \quad E_{a\uparrow} = -\Delta, \tag{2.86}$$

$$\begin{bmatrix} 0 \\ 0 \\ 0 \\ \psi_{s\downarrow\boldsymbol{k}=0} \end{bmatrix} \quad \text{with} \quad E_{s\downarrow} = \Delta. \tag{2.87}$$

The energies of these states correspond to the bottom edge of the conduction band and to the top edge of the valence band at $\boldsymbol{k} = 0$, as shown in Fig. 2.7. Comparing this figure to the gapless spectrum

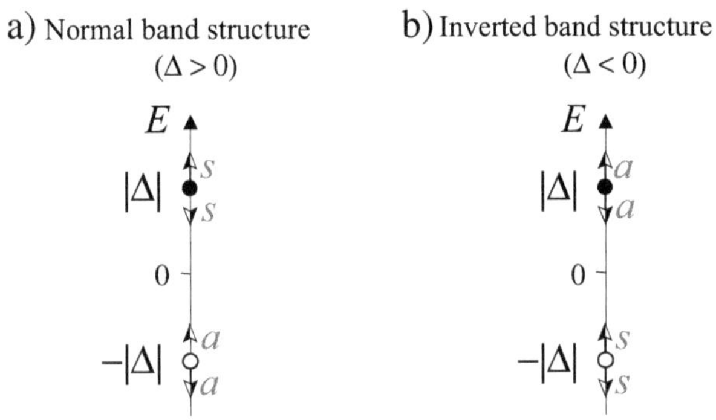

Figure 2.7 Conduction and valence band states at $\boldsymbol{k} = 0$ from Eqs. (2.84)–(2.87): (a) normal band structure for $\Delta > 0$ and (b) inverted band structure for $\Delta < 0$. s and a refer to symmetric and anti-symmetric states [see also Eq. (2.77) and text].

in Fig. 1.8, we see that now there is an energy gap $2|\Delta|$ between the conduction and valence bands. The sign of Δ is of particular importance. For $\Delta > 0$ [Fig. 2.7a] the edges of the conduction and valence bands consist of the symmetric (s) and anti-symmetric (a) states, respectively, that is, the symmetric states lie higher in energy than the anti-symmetric ones. This is the normal type of a band structure. When Δ changes its sign to negative [Fig. 2.7b], an *inverted* band structure emerges where the anti-symmetric states have a higher energy than the symmetric ones. We will see that such an unusual inverted band structure is characteristic of TIs.

In order to derive the band structure for $\boldsymbol{k} \neq 0$ we note that Eq. (2.80) decouples into two equations,

$$(\hbar v \boldsymbol{\sigma} \cdot \boldsymbol{k} + \Delta \sigma_z) \begin{bmatrix} \psi_{s\uparrow \boldsymbol{k}} \\ \psi_{a\downarrow \boldsymbol{k}} \end{bmatrix} = E \begin{bmatrix} \psi_{s\uparrow \boldsymbol{k}} \\ \psi_{a\downarrow \boldsymbol{k}} \end{bmatrix}, \tag{2.88}$$

$$-(\hbar v \boldsymbol{\sigma} \cdot \boldsymbol{k} + \Delta \sigma_z) \begin{bmatrix} \psi_{a\uparrow \boldsymbol{k}} \\ \psi_{s\downarrow \boldsymbol{k}} \end{bmatrix} = E \begin{bmatrix} \psi_{a\uparrow \boldsymbol{k}} \\ \psi_{s\downarrow \boldsymbol{k}} \end{bmatrix}, \tag{2.89}$$

which we can also write as

$$\mathrm{sgn}(\Delta)\sqrt{(\hbar v \boldsymbol{k})^2 + \Delta^2}\,(\boldsymbol{\sigma} \cdot \boldsymbol{u}) \begin{bmatrix} \psi_{s\uparrow \boldsymbol{k}} \\ \psi_{a\downarrow \boldsymbol{k}} \end{bmatrix} = E \begin{bmatrix} \psi_{s\uparrow \boldsymbol{k}} \\ \psi_{a\downarrow \boldsymbol{k}} \end{bmatrix}, \tag{2.90}$$

$$-\mathrm{sgn}(\Delta)\sqrt{(\hbar v \boldsymbol{k})^2 + \Delta^2}\,(\boldsymbol{\sigma} \cdot \boldsymbol{u}) \begin{bmatrix} \psi_{a\uparrow \boldsymbol{k}} \\ \psi_{s\downarrow \boldsymbol{k}} \end{bmatrix} = E \begin{bmatrix} \psi_{a\uparrow \boldsymbol{k}} \\ \psi_{s\downarrow \boldsymbol{k}} \end{bmatrix}, \tag{2.91}$$

where $\boldsymbol{u}$ is a unit vector defined by

$$\boldsymbol{u} = \frac{(\mathrm{sgn}(\Delta)\hbar v k_x,\ \mathrm{sgn}(\Delta)\hbar v k_x,\ |\Delta|)}{\sqrt{(\hbar v \boldsymbol{k})^2 + \Delta^2}}. \tag{2.92}$$

Each of Eqs. (2.90) and (2.91) have again two solutions being the eigenstates of the helicity matrix, $\boldsymbol{\sigma} \cdot \boldsymbol{u}$. The latter is a generalization of σ_z in Eqs. (2.82) and (2.83) and, too, has the eigenvalues equal to ± 1. Instead of Eqs. (2.84)–(2.87) we have the following four solutions

$$\begin{bmatrix} \psi_{s\uparrow \boldsymbol{k}} \\ \psi_{a\downarrow \boldsymbol{k}} \\ 0 \\ 0 \end{bmatrix}_{1,2} \quad \text{with} \quad E_{\boldsymbol{k}}^{(1,2)} = \pm\mathrm{sgn}(\Delta)\sqrt{(\hbar v \boldsymbol{k})^2 + \Delta^2}, \tag{2.93}$$

$$\begin{bmatrix} 0 \\ 0 \\ \psi_{a\uparrow \boldsymbol{k}} \\ \psi_{s\downarrow \boldsymbol{k}} \end{bmatrix}_{3,4} \quad \text{with} \quad E_{\boldsymbol{k}}^{(3,4)} = \mp\mathrm{sgn}(\Delta)\sqrt{(\hbar v \boldsymbol{k})^2 + \Delta^2}, \tag{2.94}$$

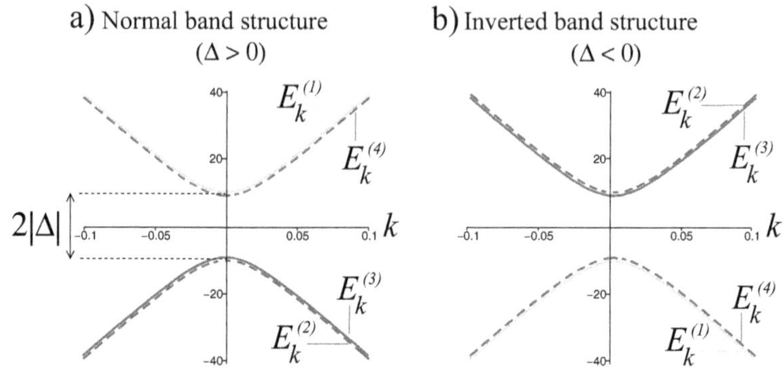

Figure 2.8 Energy bands (in meV) versus absolute value of wave vector $k = |\boldsymbol{k}|$ (in nm^{-1}) [see Eqs. (2.93) and (2.94)]: (a) normal band structure for $\Delta > 0$ and (b) inverted band structure for $\Delta < 0$; $\hbar v = 380$ meV nm and $|\Delta| = 10$ meV.

where labels 1, 2 and 3, 4 correspond to the signs $\pm$ and $\mp$, respectively. We see that for $\boldsymbol{k} \neq 0$ the symmetric and anti-symmetric states are mixed. For instance, solution 1 coincides at $\boldsymbol{k} = 0$ with the symmetric state $s \uparrow$ [see Eq. (2.84)], acquiring for $\boldsymbol{k} \neq 0$ an admixture of the anti-symmetric state $a \downarrow$ [see Eq. (2.85)]. The corresponding band dispersions are shown in Figs. 2.8 and 2.9.

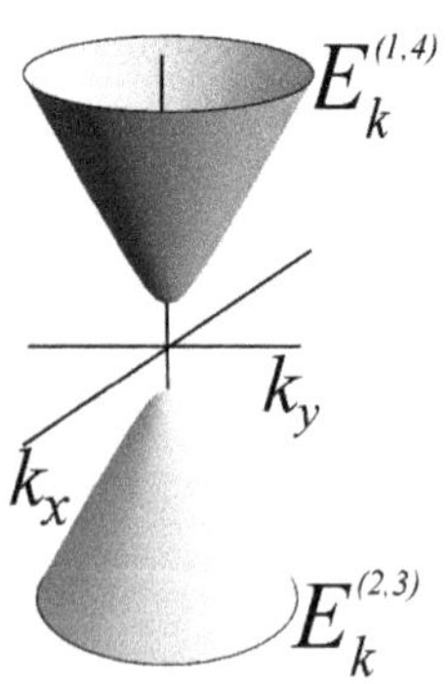

Figure 2.9 3D plot of energy bands in Eqs. (2.93) and (2.94) for $\hbar v = 380$ meV nm and $\Delta = 10$ meV. Note that conduction and valence bands consist of degenerate branches 1, 4 and 2, 3, respectively.

2.2.3 *Edge States. Reduction of Bulk Degrees of Freedom and Spin Helicity*

We are now in a position to examine the topological properties of the system by looking at its edge excitations. The boundary problem that we intend to solve is given by Eqs. (2.62), (2.64) and (2.71). To write them in a more compact form we introduce another set of Pauli matrices $\tau_{x,y}$ and τ_z that operate in the space of the t and b subsystems. Then, the boundary problem assumes the form

$$\left[-i\hbar v\tau_z(\sigma_x\partial_x + \sigma_y\partial_y) + \Delta\tau_x\sigma_0\right]\psi(\mathbf{r}) = E\psi(\mathbf{r}), \tag{2.95}$$

$$\psi(x, y = 0) = \tau_y\sigma_y\psi(x, y = 0). \tag{2.96}$$

We again seek the solutions that propagate along the edge with a wave vector k and decay exponentially in the bulk on a length λ:

$$\psi_k(x, y) = \Phi\, e^{ikx - y/\lambda}, \tag{2.97}$$

$$\Phi = \tau_y\sigma_y\Phi. \tag{2.98}$$

The constraint (2.98) on the spinor Φ follows from the boundary condition (2.96). Inserting Eq. (2.97) into Eq. (2.95) and using the identity $\tau_z\sigma_y\Phi = -i\tau_x\sigma_0\Phi$, we have

$$\hbar v k\tau_z\sigma_x\Phi + (\Delta + \hbar v/\lambda)\,\tau_x\sigma_0\Phi = E\Phi. \tag{2.99}$$

Setting then the expression in the brackets to zero, we find the decay length

$$\lambda = -\hbar v/\Delta, \tag{2.100}$$

while Eq. (2.99) yields

$$\hbar v k\tau_z\sigma_x\Phi = E\Phi. \tag{2.101}$$

We are about to make an important observation: At the edge the number of independent degrees of freedom is reduced to the electron spin only. Indeed, the constraint (2.96) restricts the form of the spinor Φ to

$$\Phi = \begin{bmatrix} \chi \\ i\sigma_y\chi \end{bmatrix}, \tag{2.102}$$

due to which Eq. (2.101) reduces to an equation for the two-component spinor χ,

$$\hbar v k\sigma_x\chi = E\chi. \tag{2.103}$$

Its solutions are the eigenstates of the spin projection σ_x,

$$\sigma_x \chi_s = s\chi_s, \quad s = \pm 1, \tag{2.104}$$

with the energies

$$E_{ks} = \hbar v k s. \tag{2.105}$$

Finally, the normalized edge states (2.97) are

$$\psi_{k+} = \begin{bmatrix} 1 \\ 1 \\ 1 \\ -1 \end{bmatrix} \sqrt{\frac{|\Delta|}{2\hbar v}}\, e^{ikx-|\Delta|y/\hbar v}, \tag{2.106}$$

and

$$\psi_{k-} = \begin{bmatrix} 1 \\ -1 \\ -1 \\ -1 \end{bmatrix} \sqrt{\frac{|\Delta|}{2\hbar v}}\, e^{ikx-|\Delta|y/\hbar v}. \tag{2.107}$$

These results illustrate all the hallmarks of the TR-invariant topological insulators, as summarized below:

- The edge states are gapless, which can again be attributed to the inversion of the band structure at the edge. Indeed, the system can be extended to the other semispace ($y < 0$) by the parity transformation,

  $$\psi'(x, y) = \tau_y \sigma_y \psi(x, -y), \tag{2.108}$$

  that inverts the sign of the gap term in Eq. (2.95)

  $$[-i\hbar v \tau_z(\sigma_x \partial_x + \sigma_y \partial_y) - \Delta \tau_x \sigma_0]\psi'(x, y) = E\, \psi'(x, y), \tag{2.109}$$

  so that the boundary condition (2.96) is equivalent to a gap domain wall at the edge. Inside the physical area, the band structure is inverted, which is a typical situation in TI materials, for example, in HgTe quantum wells.
- The bulk degrees of freedom are reduced at the edge to the electron spin only. Unlike the Chern insulator, there are two counter-propagating edge states with the opposite spin projections σ_x (see also Fig. 2.10). That is, the propagation direction of the electron is locked to the orientation of

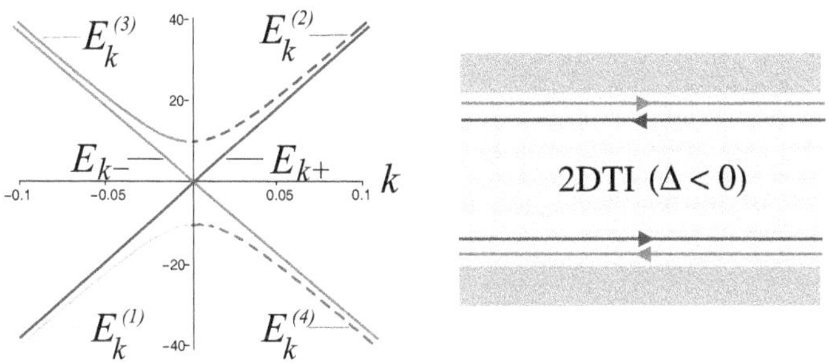

Figure 2.10 Band structure of a 2DTI with gapless helical edge states, $E_{k\pm}$, [see Eq. (2.105)] for $\hbar v = 380$ meV nm and $\Delta = -10$ meV (left) and schematic of helical (counter-propagating) edge states at opposite boundaries of a 2DTI (right).

its spin. This property is known as helicity, due to which the edge states in 2DTIs are frequently called helical, as opposed to the chiral edge states in the quantum Hall systems. The helical edge states are related to each other by the TRS,

$$\psi_{-k,-} = \mathbb{T}\,\psi_{k,+}, \tag{2.110}$$

where $\mathbb{T}$ is defined in Eq. (2.65). Also, in accord with Kramers' theorem and Eqs. (1.13) and (1.16), the edge states (2.106) and (2.107) are orthogonal, and hence, protected against local perturbations that do not break the TRS. One of the most common types of such perturbations in realistic systems is potential disorder created by randomly distributed ionized impurities. If the impurities do not flip the spin of the edge state, its propagation direction remains unchanged, which means that potential disorder cannot cause scattering between the helical edge channels and their localization. Such robustness of the edge states has important implications for electron transport in 2DTIs.

2.2.4 Z_2 Invariant

In view of the TRS, the Hall conductivity in 2DTIs is zero. This means that the TKNN topological invariant—the first Chern number (2.61)—vanishes. Still, it is possible to define a topological invariant, the so called Z_2 invariant [2], that distinguishes between the topologically trivial and nontrivial phases of a TR-invariant system. Let us consider, for example, the representation (2.78) of the TI Hamiltonian obtained after the unitary transformation (2.76). Since the transformed Hamiltonian is block diagonal, the Hall conductivity (2.59) is

$$\sigma_{xy} = \frac{e^2}{h}(n_+ + n_-), \tag{2.111}$$

where the numbers $n_\pm$ are given by Eq. (2.61) for the upper and lower blocks of Eq. (2.78). By TRS the gap has opposite signs in the two blocks, which yields

$$n_+ + n_- = 0. \tag{2.112}$$

The opposite signs of the TKNN numbers $n_\pm$ correspond to the opposite chiralities of the helical edge states. The sum $n_+ + n_-$ is related to the (vanishing) total charge carried by the pair. On the other hand, the difference $n_+ - n_-$ can be linked to the total spin accumulated by the Kramers pair at the edge [1]. A topological invariant defined as

$$\frac{n_+ - n_-}{2} \tag{2.113}$$

appears to be associated with a quantized spin-Hall conductivity [1]. In this particular case, the Z_2 index (2.113) is identical to the spin invariant $\tau_z\sigma_0$ of the Hamiltonian (2.78). A more general approach to the Z_2 invariant adopted in recent literature exploits the symmetry of the matrix elements of the TR operator in the Brillouin zone [2]. Further, details of this theoretical framework have been reviewed in Refs. [18, 19, 52].

2.2.5 *HgTe Quantum Wells. Nonlocal Edge Transport*

The helical edge states in 2DTIs generate nonlocal transport effects that have no analogues in conventional 2D systems. This has been observed, for the first time, in experiments on HgTe quantum wells reported in Ref. [7]. HgTe has a zinc blende lattice type, shown in Fig. 2.11, with the first Brillouin zone in the form of an octahedron. The conduction and valence bands meet at the center of the Brillouin zone and have the s- and p-type symmetries inherited from the

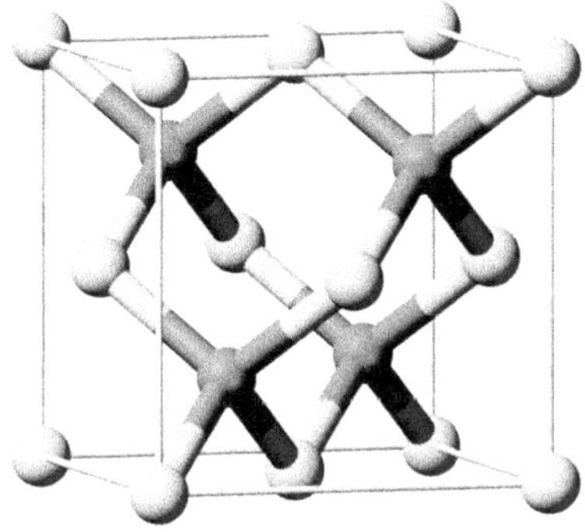

Figure 2.11 Zinc blende lattice structure of HgTe. Hg and Te atoms are depicted as dark blue and yellow spheres, respectively.

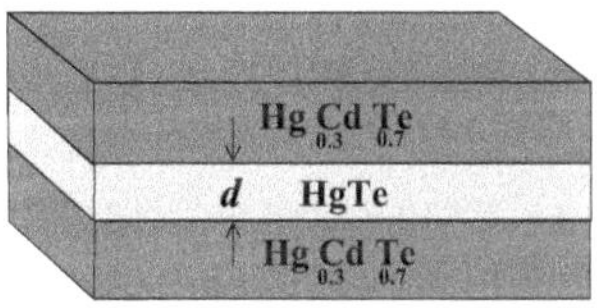

Figure 2.12 Schematic of HgTe quantum well structure. The HgTe layer hosts a 2D system that exhibits the 2DTI (quantum spin-Hall) state.

atomic orbitals of Hg and Te, respectively. In the presence of the crystal potential, the bands are split by the $\boldsymbol{k} \cdot \boldsymbol{p}$ and spin–orbit couplings (Ref. [53]), resulting in the Dirac-like dispersion of HgTe quantum wells [4, 54], similar to that described by Hamiltonian (2.78). Moreover, because of the atomically heavy mercury the relativistic corrections to the band structure are large, leading to the inverted band structure where the s-type band (usually acting as the conduction band) has a lower energy than the p-type ones. Thus, the conditions needed for the formation of the helical edge states can be met in appropriately designed HgTe quantum well structures. A sketch of such a structure is shown in Fig. 2.12 (Ref. [6]).

In order to illustrate the nonlocal transport through the helical edge states we follow Ref. [7] and consider a four-terminal Hall bar depicted in Fig. 2.13. For comparison, panel a of Fig. 2.13 shows a chiral edge state in a strong perpendicular magnetic field (*cf.* Fig. 1.1 of Chapter 1), whereas panel b shows the helical edge states

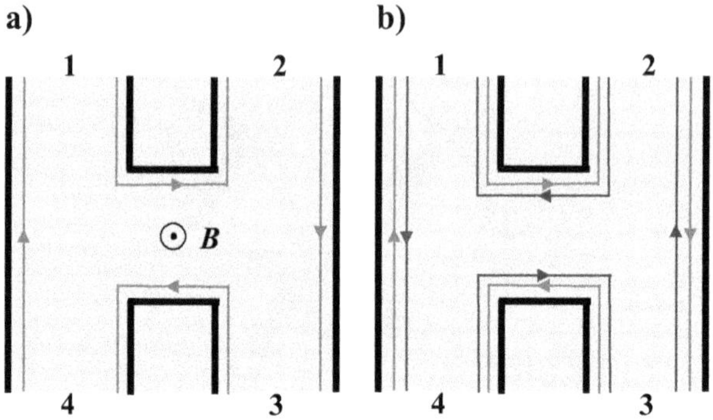

Figure 2.13 Schematic of edge transport in four-terminal device: (a) chiral edge state in a strong magnetic field $\boldsymbol{B}$ and (b) helical edge states in a 2DTI in zero magnetic field.

in a 2DTI in a zero magnetic field. The Hall bar terminals serve to inject electric currents and (or) to measure induced voltages. We are interested in the relation between current I_i injected through contact i and voltages V_j induced on all four contacts (i, $j = 1.4$). For that purpose, one commonly uses the Landauer–Buetikker formula,

$$I_i = \frac{e^2}{h} \sum_{j=1}^{N} (T_{ji} V_i - T_{ij} V_j), \tag{2.114}$$

where T_{ji} is the transmission probability from contact i to contact j. Let us consider the chiral edge state in Fig. 2.13a. Since it connects the neighboring contacts in one propagation direction, the corresponding transmission coefficient, T_{ji}^{chi}, can be written as

$$T_{i+1,i}^{chi} = 1, \qquad i = 1, \ldots, N, \tag{2.115}$$

where N is the number of the terminals (i.e., $N = 4$ in Fig. 2.13), with the convention that $T_{N+1,N} = T_{1,N}$ describes the transmission from terminal N to terminal 1. Assuming, for concreteness, that the current flows from terminal 1 to terminal 4, while leads 2 and 3 are used as voltage probes, we have from Eq. (2.114)

$$I_1 \equiv I_{14} = \frac{e^2}{h}(V_1 - V_4), \;\; V_2 = V_1, \;\; V_3 = V_2, \;\; I_4 = -I_1. \tag{2.116}$$

This yields a finite two-terminal resistance

$$R_{14,14} = \frac{V_1 - V_4}{I_{14}} = \frac{h}{e^2}, \tag{2.117}$$

and zero four-terminal (nonlocal) resistances

$$R_{14,12} = \frac{V_1 - V_2}{I_{14}} = 0, \tag{2.118}$$

$$R_{14,23} = \frac{V_2 - V_3}{I_{14}} = 0, \tag{2.119}$$

$$R_{14,13} = \frac{V_1 - V_3}{I_{14}} = 0. \tag{2.120}$$

In contrast, the helical edge channels connect the neighboring contacts in both propagation directions with the transmission coefficients, T_{ji}^{hel}, given by

$$T_{i+1,i}^{hel} = T_{i,i+1}^{hel} = 1, \qquad i = 1, \ldots, N, \tag{2.121}$$

with the conventions $T_{N+1,N} = T_{1,N}$ and $T_{N,N+1} = T_{N,1}$. As the helical edge states are protected by the TRS, the transmission between the contacts is perfect even in the presence of potential disorder (e.g., ionized impurities). For the same current contacts (1 and 4) and with $T_{i+1,i}^{hel}$ given by Eq. (2.121), we find from Eq. (2.114)

$$I_1 \equiv I_{14} = \frac{e^2}{h}(2V_1 - V_4 - V_2), \qquad I_4 = -I_1, \tag{2.122}$$

$$2V_2 - V_1 - V_3 = 0, \qquad 2V_3 - V_2 - V_4 = 0. \tag{2.123}$$

This, in turn, yields the two-terminal resistance [7]

$$R_{14,14} = \frac{V_1 - V_4}{I_{14}} = \frac{3}{4}\frac{h}{e^2}, \tag{2.124}$$

and the four-terminal resistances [7]

$$R_{14,12} = \frac{V_1 - V_2}{I_{14}} = R_{14,23} = \frac{V_2 - V_3}{I_{14}} = \frac{1}{4}\frac{h}{e^2}, \tag{2.125}$$

$$R_{14,13} = \frac{V_1 - V_3}{I_{14}} = \frac{1}{2}\frac{h}{e^2}. \tag{2.126}$$

We see that the helical transport is characterized by the lower two-terminal resistance $R_{14,14}$ [Eq. (2.124)]. However, the most striking feature of the helical edge states is the nonzero nonlocal resistances (2.125) and (2.126). They are the direct consequence of the TRS and are unique to the 2DTIs [7].

2.2.6 Related Research on 2DTI Materials

The discovery of the protected helical edge states and nonlocal transport in 2DTIs has stimulated vigorous interest in their transport properties. Some of them have been reviewed in Refs. [6, 18, 19, 55, 56]. This includes a variety of transport phenomena specific to 2DTIs and HgTe quantum wells that have been studied theoretically in Refs. [57–91, 93–107, 122, 158, 159]. Among these, Ref. [59] has shown that the 2DTI state disappears in narrow wires due to an overlap of the edge states from the opposite sample boundaries. A number of papers have addressed the issue of disorder in 2DTIs (Refs. [65–67, 82, 84, 94, 96, 97, 105]). In particular, Refs. [66] and [67] have proposed microscopic mechanisms of the magnetoresistance in disordered 2DTIs subject to an external magnetic field. Another issue is electron interactions in the helical edge channels (Refs. [68, 72, 76, 89, 91, 100, 102, 103, 107]). It has been found that the electron backscattering may occur if the axial spin rotation symmetry of Hamiltonian (2.78) is broken, for example, by the Bychkov–Rashba spin–orbit coupling [68, 91, 102]. The spatially random Bychkov–Rashba coupling has been predicted to cause localization of the edge states in the presence of weakly screened electron–electron interaction [68]. Reference [102] has shown that weak electron–electron interaction in the absence of the axial spin rotation symmetry allows for inelastic backscattering of a single electron, accompanied by forward scattering of another. Of particular interest are magnetic properties of 2DTIs. These have been addressed in several publications (Refs. [60, 98, 99, 104]). Theoretical aspects of electron correlation effects in 2DTIs have been reviewed in Ref. [106].

As to the recent experiments on 2DTIs, one should mention the measurements of spin polarization (Ref. [108]) and backscattering (Ref. [109]) of the helical edge states in HgTe quantum wells in the quantum spin-Hall regime. In other experiments on HgTe quantum wells (Ref. [110]), a large nonlocal electric response near the charge neutrality point in the presence of a perpendicular magnetic field has been observed. Also, helical edge modes and the quantum spin-Hall state have been reported to occur in inverted InAs/GaSb quantum wells [113, 114].

2.3 Problems

(1) Prove that in the transformed basis (2.77) the boundary condition (2.96) takes the form

$$\psi'(x, y = 0) = \hat{M}'\psi'(x, y = 0), \qquad \hat{M}' = \begin{bmatrix} \sigma_x & 0 \\ 0 & \sigma_x \end{bmatrix}. \quad (2.127)$$

(2) Consider the boundary problem given by Eqs. (2.73), (2.78), and (2.127). Find the edge solutions and show that they are helical with the spectrum given by Eq. (2.105). Identify further properties of these edge states that are independent of the basis.

(3) In Ref. [4] Bernevig, Hughes, and Zhang (BHZ) derive the following Hamiltonian for HgTe quantum wells:

$$\hat{H}_{BHZ} = \begin{bmatrix} h(\boldsymbol{k}) & 0 \\ 0 & h^*(-\boldsymbol{k}) \end{bmatrix}, \quad (2.128)$$

$$h(\boldsymbol{k}) = \mathcal{A}(\sigma_x k_x - \sigma_y k_y) + \mathcal{M}_{\boldsymbol{k}}\sigma_z + \mathcal{D}\boldsymbol{k}^2\sigma_0, \quad (2.129)$$

$$\mathcal{M}_{\mathbf{k}} = \mathcal{M} + \mathcal{B}\mathbf{k}^2, \quad (2.130)$$

where $\mathcal{A}$, $\mathcal{B}$, $\mathcal{D}$, and $\mathcal{M}$ are constants characterizing the band structure. Using the unitary transformation

$$U = \begin{bmatrix} 0 & \sigma_z \\ -i\sigma_y & 0 \end{bmatrix}, \quad (2.131)$$

obtain the Hamiltonian $\hat{H}'_{BHZ} = U\hat{H}_{BHZ}U^{\dagger}$. Show that for $\mathcal{D} = \mathcal{B} = 0$ the new Hamiltonian coincides with $\hat{H}'$ in Eq. (2.78).

Chapter 3

Two-Dimensional Topological Insulators in Strong Magnetic Fields

We have seen that the helical edge states in 2DTIs form a Kramers doublet with the two states being related by the TRS. If a strong external magnetic field is applied, the TRS is no longer a valid symmetry. What happens to the helical edge states in that case? Here we address this question, which is of both theoretical and practical interest. The answer turns out to be quite counter-intuitive: the helical edge states can still exist in strong quantizing magnetic fields. The robustness of the 2DTIs is the consequence of the fact that the edge states are bound to an effective gap domain wall that is not destroyed by a magnetic field. However, as we shall see here, both spectral and transport properties of the helical edge states are modified by a strong quantizing magnetic field [67, 75, 94].

3.1 Helical Edge States in a Magnetic Field

3.1.1 *Boundary Problem. Green Function Method*

We consider the same geometry as in Section 2.2.1, but now in the presence of a perpendicular magnetic field $\boldsymbol{B}$ as shown in Fig. 3.1.

Topological Insulators: The Physics of Spin Helicity in Quantum Transport
Grigory Tkachov

ISBN 978-981-4613-25-5 (Hardcover), 978-981-4613-26-2 (eBook)
www.panstanford.com

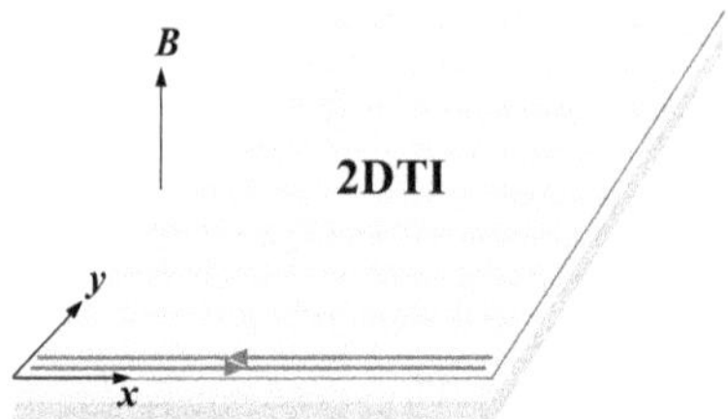

Figure 3.1 Geometry of a 2DTI with helical edge states at $y = 0$ in a perpendicular magnetic field $\boldsymbol{B}$.

The gauge of the vector potential $\boldsymbol{A}(\boldsymbol{r})$ is chosen to be

$$\boldsymbol{A}(\boldsymbol{r}) = (-By, 0, 0). \tag{3.1}$$

The boundary problem that we intend to solve is a generalization of that considered in Chapter 2 [see also Eqs. (2.73), (2.78), and (2.127)],

$$\begin{bmatrix} E\sigma_0 - v\boldsymbol{\sigma}\cdot\boldsymbol{P} - \Delta\sigma_z & 0 \\ 0 & E\sigma_0 + v\boldsymbol{\sigma}\cdot\boldsymbol{P} + \Delta\sigma_z \end{bmatrix} \hat{G}(\boldsymbol{r}, \boldsymbol{r}') = \delta(\boldsymbol{r} - \boldsymbol{r}'), \tag{3.2}$$

$$\boldsymbol{P} = -i\hbar\nabla - \frac{e}{c}\boldsymbol{A}(\boldsymbol{r}), \tag{3.3}$$

$$\hat{G}(\boldsymbol{r}, \boldsymbol{r}')|_{y=0} = \begin{bmatrix} \sigma_x & 0 \\ 0 & \sigma_x \end{bmatrix} \hat{G}(\boldsymbol{r}, \boldsymbol{r}')|_{y=0}. \tag{3.4}$$

Instead of the spinor wave function, in this chapter we work with the matrix Green function $\hat{G}(\boldsymbol{r}, \boldsymbol{r}')$. This has an extra goal to acquaint the reader with the Green function method, which will be needed in the subsequent chapters. Concretely, we consider the retarded Green function with a positive infinitesimally small shift of the energy, that is $E \to E + i0$.

Unlike the equation for the wave function, Eq. (3.2) contains a delta-functional source term, $\delta(\boldsymbol{r} - \boldsymbol{r}')$, on the right-hand side. The Green function $\hat{G}(\boldsymbol{r}, \boldsymbol{r}')$ is a 4×4 block-diagonal matrix of the form

$$\hat{G}(\boldsymbol{r}, \boldsymbol{r}') = \begin{bmatrix} \hat{G}_+(\boldsymbol{r}, \boldsymbol{r}') & 0 \\ 0 & \hat{G}_-(\boldsymbol{r}, \boldsymbol{r}') \end{bmatrix}, \tag{3.5}$$

where $\hat{G}_+(\boldsymbol{r}, \boldsymbol{r}')$ is a 2×2 matrix in the basis of spin states $s \uparrow$ and $a \downarrow$ [*cf.* upper two spinor components in Eq. (2.77)], whereas $\hat{G}_-(\boldsymbol{r}, \boldsymbol{r}')$

is a 2 × 2 matrix in the basis of spin states a ↑ and d ↓ [*cf.* lower two spinor components in Eq. (2.77)]. Using again the block index $\tau = \pm$ (or $\tau = \pm 1$ as a variable in algebraic expressions) we can write $\hat{G}_{\pm}(\boldsymbol{r}, \boldsymbol{r}')$ as

$$\hat{G}_{\tau}(\boldsymbol{r}, \boldsymbol{r}') = \begin{bmatrix} G_{\uparrow\uparrow,\tau}(\boldsymbol{r}, \boldsymbol{r}') & G_{\uparrow\downarrow,\tau}(\boldsymbol{r}, \boldsymbol{r}') \\ G_{\downarrow\uparrow,\tau}(\boldsymbol{r}, \boldsymbol{r}') & G_{\downarrow\downarrow,\tau}(\boldsymbol{r}, \boldsymbol{r}') \end{bmatrix}. \tag{3.6}$$

Writing Eq. (3.2) in components, it is easy to express the off-diagonal entries $G_{\uparrow\downarrow,\tau}$ and $G_{\downarrow\uparrow,\tau}$ in terms of the diagonal ones as

$$G_{\uparrow\downarrow,\tau}(\boldsymbol{r}, \boldsymbol{r}') = \frac{v(P_x - iP_y)}{\tau E - \Delta} G_{\downarrow\downarrow,\tau}(\boldsymbol{r}, \boldsymbol{r}'), \tag{3.7}$$

$$G_{\downarrow\uparrow,\tau}(\boldsymbol{r}, \boldsymbol{r}') = \frac{v(P_x + iP_y)}{\tau E + \Delta} G_{\uparrow\uparrow,\tau}(\boldsymbol{r}, \boldsymbol{r}'), \tag{3.8}$$

where $G_{\uparrow\uparrow,\tau}$ and $G_{\downarrow\downarrow,\tau}$ obey the equations

$$[E^2 - \Delta^2 - v^2(P_x - iP_y)(P_x + iP_y)]G_{\uparrow\uparrow,\tau} = (E + \Delta\tau)\delta(\mathbf{r} - \mathbf{r}'), \tag{3.9}$$

$$[E^2 - \Delta^2 - v^2(P_x + iP_y)(P_x - iP_y)]G_{\downarrow\downarrow,\tau} = (E - \Delta\tau)\delta(\mathbf{r} - \mathbf{r}'). \tag{3.10}$$

For a nonzero magnetic field the momentum operators P_x and P_y do not commute (see Problem 1 at the end of this chapter). Taking this into account and making the plane-wave expansion,

$$\hat{G}_{\tau}(\boldsymbol{r}, \boldsymbol{r}') - \sum_k \hat{G}_{\tau,k}(y, y') e^{ik(x-x')}/L, \tag{3.11}$$

where L is the system length in the x-direction, we reduce Eqs. (3.9) and (3.10) to a pair of ordinary differential equations for $G_{\uparrow\uparrow,\tau,k}(y, y')$ and $G_{\downarrow\downarrow,\tau,k}(y, y')$. It is convenient to write them in the dimensionless form

$$\left[\partial_z^2 - \frac{(z-q)^2}{4} - a\right] G_{\uparrow\uparrow,\tau,k} = \frac{(E + \Delta\tau)\ell}{(\hbar v)^2}\delta(z - z'), \tag{3.12}$$

with

$$a = \frac{(\Delta^2 - E^2)\ell^2}{(\hbar v)^2} - \frac{\text{sgn}(eB)}{2}, \tag{3.13}$$

and

$$\left[\partial_z^2 - \frac{(z-q)^2}{4} - \overline{a}\right] G_{\downarrow\downarrow,\tau,k} = \frac{(E - \Delta\tau)\ell}{(\hbar v)^2}\delta(z - z'), \tag{3.14}$$

with

$$\bar{a} = \frac{(\Delta^2 - E^2)\ell^2}{(\hbar v)^2} + \frac{\operatorname{sgn}(eB)}{2}. \tag{3.15}$$

In Eqs. (3.12) and (3.14),

$$z = \frac{y}{\ell} \tag{3.16}$$

is the dimensionless position measured in units of the magnetic length,

$$\ell = \sqrt{\frac{c\hbar}{2|eB|}}, \tag{3.17}$$

and

$$q = -2k\ell \operatorname{sgn}(eB) \tag{3.18}$$

is the dimensionless wave vector.

The next step is to obtain the boundary conditions to Eqs. (3.12) and (3.14) from Eq. (3.4). The latter couples the off-diagonal and diagonal elements of matrix $\hat{G}_{\tau,k}$:

$$G_{\downarrow\uparrow,\tau,k}(0, z') = G_{\uparrow\uparrow,\tau,k}(0, z'), \quad G_{\uparrow\downarrow,\tau,k}(0, z') = G_{\downarrow\downarrow,\tau,k}(0, z'). \tag{3.19}$$

Eliminating the off-diagonal matrix elements with the help of Eqs. (3.7) and (3.8), we find the required boundary conditions for the diagonal entries:

$$\partial_z G_{\uparrow\uparrow,\tau,k}(z, z')\big|_{z=0} = \kappa\, G_{\uparrow\uparrow,\tau,k}(0, z'), \tag{3.20}$$

with

$$\kappa = \frac{(\tau E + \Delta)\ell}{\hbar v} + \frac{q \operatorname{sgn}(eB)}{2}, \tag{3.21}$$

and

$$\partial_z G_{\downarrow\downarrow,\tau,k}(z, z')\big|_{z=0} = \bar{\kappa}\, G_{\downarrow\downarrow,\tau,k}(0, z'), \tag{3.22}$$

with

$$\bar{\kappa} = \frac{(-\tau E + \Delta)\ell}{\hbar v} - \frac{q \operatorname{sgn}(eB)}{2}. \tag{3.23}$$

3.1.2 *Solution in Terms of Parabolic Cylinder Functions*

The boundary problem formulated above can be solved exactly in terms of the parabolic cylinder functions. It is an interesting example of such higher transcendental functions finding application in the physics of topological insulators. From this point of view, it is instructive to present here the details of the calculations. We start with Eq. (3.12) and boundary condition (3.20). For simplicity, the Green function indices $\uparrow$, τ, k are suppressed below.

The general solution of Eq. (3.12) has the form of a linear combination:

$$G(z, z') = A(z')U(a, z-q) + B(z')U(a, -z+q) \tag{3.24}$$

$$+ G^{\infty}(z, z'). \tag{3.25}$$

The sum of the first and second terms is the general solution of Eq. (3.12) with the zero right-hand side, where $U(a, z-q)$ and $U(a, -z+q)$ are the parabolic cylinder functions representing two linearly independent (fundamental) solutions of the equation [115],

$$\left[\partial_z^2 - \frac{(z-q)^2}{4} - a\right] U = 0. \tag{3.26}$$

The coefficients $A(z')$ and $B(z')$ should be found from the boundary conditions at $z \to +\infty$ and $z = 0$. The function $U(a, -z+q)$ is unbounded for $z \to +\infty$, whereas $U(a, z-q)$ vanishes in this limit [115]. Therefore, $B(z')$ in Eq. (3.24) must be set to zero

$$B(z') = 0. \tag{3.27}$$

As long as the last term in Eq. (3.25) is specified, the remaining coefficient, $A(z')$, can be found from boundary condition (3.20).

The last term in Eq. (3.25) is a particular solution of the inhomogeneous Eq. (3.12), that is, the solution generated by the nonzero right-hand side. Such a particular solution can always be chosen in the form of the Green function of the infinite system. The latter can be constructed from the two linearly independent solutions, $U(a, z-q)$ and $U(a, -z+q)$, according to the standard recipe:

$$G^{\infty}(z, z') = C\,[\,\Theta(z-z')\,U(a, z-q)U(a, -z'+q) \tag{3.28}$$

$$+ \Theta(z'-z)\,U(a, z'-q)U(a, -z+q)], \tag{3.29}$$

$$C = \frac{(E + \Delta\tau)\ell}{\hbar^2 v^2 W}, \tag{3.30}$$

$$W = \partial_z U(a, \overline{z}) \cdot U(a, -\overline{z}) - U(a, \overline{z}) \cdot \partial_z U(a, -\overline{z}) \tag{3.31}$$

$$= -\frac{\sqrt{2\pi}}{\Gamma(a + 1/2)}, \qquad \overline{z} = z - q, \tag{3.32}$$

where $\Theta(z)$ is the unit-step function, W is the Wronskian of the parabolic cylinder functions [115], and $\Gamma(a + 1/2)$ is Euler's gamma function. We note that the constructed solution $G^{\infty}(z, z')$ vanishes for $z \to \pm\infty$.

Inserting Eq. (3.24) into boundary condition (3.20), we find the coefficient $A(z')$ as

$$A(z') = -C \frac{\partial_q U(a, q) + \kappa U(a, q)}{\partial_q U(a, -q) + \kappa U(a, -q)} U(a, z' - q). \tag{3.33}$$

Inserting this back into Eq. (3.24), we obtain the explicit solution of our boundary problem,

$$G(z, z') = G^{\infty}(z, z') - C \frac{\partial_q U(a, q) + \kappa U(a, q)}{\partial_q U(a, -q) + \kappa U(a, -q)} \tag{3.34}$$

$$\times\, U(a, z - q) U(a, z' - q). \tag{3.35}$$

The helical edge states correspond to the poles of the second term, that is, the edge-state spectrum is obtained from the equation:

$$\partial_q U[a(E), -q] + \kappa(E) U[a(E), -q] = 0, \tag{3.36}$$

in which $a(E)$ and $\kappa(E)$ are the known functions of energy [see Eqs. (3.13) and (3.21)].

For our further discussion, we only need the asymptotic formulas with energy E close to the zeros of Eq. (3.36). For that purpose, we rewrite Eqs. (3.34) and (3.35) in a slightly different form,

$$G(z, z') = G^{\infty}(z, z') \tag{3.37}$$

$$- C \frac{U(a, q)}{U(a, -q)} U(a, z - q) U(a, z' - q) \tag{3.38}$$

$$- CW \frac{U(a, z - q) U(a, z' - q)}{U(a, -q)[\partial_q U(a, -q) + \kappa U(a, -q)]}, \tag{3.39}$$

which near the poles of the last (edge) term reduces to

$$G(z, z') \approx -CW \frac{U(a, z-q)U(a, z'-q)}{U(a, -q)[\partial_q U(a, -q) + \kappa U(a, -q)]}. \tag{3.40}$$

Next, we use the recurrence relation for the parabolic cylinder functions [115],

$$\partial_q U(a, -q) = -\frac{q}{2} U(a, -q) + \left(a + \frac{1}{2}\right) U(a+1, -q), \tag{3.41}$$

to eliminate the derivative in the denominator of Eq. (3.40). After simple algebra involving Eqs. (3.13), (3.21), and (3.31) and assuming for concreteness $eB > 0$, we arrive at the final result:

$$G_{\uparrow\uparrow,\tau,k}(z, z') \approx \frac{U(a, z-q)U(a, z'-q)}{\ell\, U(a, -q)U(a+1, -q)} \tag{3.42}$$

$$\times \frac{1}{E - \Delta\tau - (\hbar v/\ell)U(a, -q)/U(a+1, -q)}, \tag{3.43}$$

from which we obtain the equation for the edge-state spectrum as

$$E - \Delta\tau - \frac{\hbar v}{\ell} \frac{U[a(E), -q]}{U[a(E)+1, -q]} = 0. \tag{3.44}$$

In Eq. (3.43) we re-installed the indices of the Green function to distinguish it from the other diagonal Green function, $G_{\downarrow\downarrow,\tau,k}(z, z')$. It can be obtained from Eqs. (3.14) and (3.22) in a fully analogous way. Here we quote the final result

$$G_{\downarrow\downarrow,\tau,k}(z, z') \approx \frac{U(a+1, z-q)U(a+1, z'-q)U(a, -q)}{\ell\, U^3(a+1, -q)} \tag{3.45}$$

$$\times \frac{1}{E - \Delta\tau - (\hbar v/\ell)U(a, -q)/U(a+1, -q)}, \tag{3.46}$$

These results are valid for all values of the parameter $\hbar v/\ell|\Delta|$ that quantifies the relative strength of the magnetic field, B. In particular, the limits

$$\frac{\hbar v}{\ell|\Delta|} \ll 1, \tag{3.47}$$

and

$$\frac{\hbar v}{\ell|\Delta|} \gg 1, \tag{3.48}$$

correspond to the *weak*- and *strong*-field regimes, respectively. In the weak-field regime, the results obtained here recover the

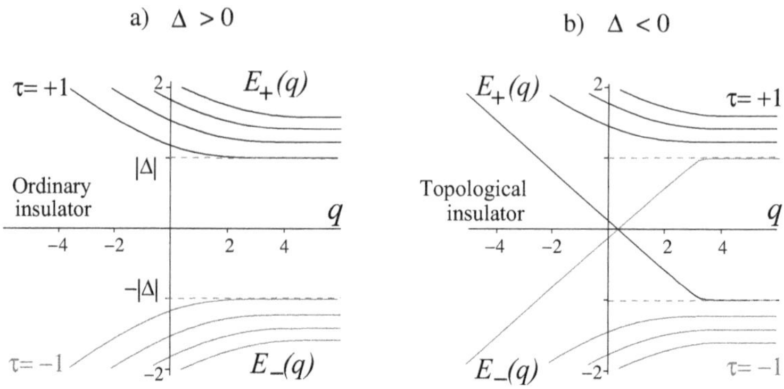

Figure 3.2 Edge-state energies $E_+(q)$ and $E_-(q)$ as function of dimensionless wave vector, q [see Eqs. (3.18) and (3.44)] for (a) normal band structure with $\Delta > 0$ and (b) inverted band structure with $\Delta < 0$; $\hbar v/\ell|\Delta| = 0.5$ and $eB > 0$. Panel b shows a 2DTI state with two helical (counter-propagating) modes crossing inside the gap. At large $q \gg 1$ helical edge states evolve into flat bulk Landau levels $\pm|\Delta|$.

edge-state solution discussed in Chapter 2 (Refs. [67, 94]). Below we focus on the helical edge states in strong quantizing fields.

3.1.3 *Nonlinear Helical Spectrum*

In order to obtain the edge-state spectrum we solve numerically, Eq. (3.44) for energy E as function of dimensionless momentum q for both $\tau = \pm 1$. The numerical solutions $E_+(q)$ and $E_-(q)$ are shown in Fig. 3.2. Again, the spectrum depends crucially on the sign of the gap parameter Δ. The system with the normal band structure ($\Delta > 0$) remains an ordinary band insulator at subgap energies $|E| < \Delta$, as shown in Fig. 3.2a. Outside the band gap the conduction- and valence-band states are quantized into discrete edge-state branches evolving into flat bulk Landau levels with increase in $q > 0$. Since these states are chiral (i.e., propagating in the same direction), the TRS is broken.

The results for the inverted band structure with $\Delta < 0$ are qualitatively different (see Fig. 3.2b). Apart from the quantized conduction- and valence-band states, there are two counter-propagating edge modes crossing in the middle of the band gap.

These are the high-field counterparts of the helical edge states studied in Chapter 2. Thus, we conclude that, despite the broken TRS in the conduction and valence bands, the 2DTI state persists in strong fields. This agrees with the qualitative picture of the helical states as robust domain wall states (see also Fig. 3.1b).

In the strong-field regime (3.48), an analytical formula for the edge-state spectrum can be derived [67]. Using the asymptotic expressions (see Ref. [115])

$$U(a, z) \approx U(-1/2, z) = e^{-z^2/4}, \tag{3.49}$$

$$U(a+1, z) \approx U(1/2, z) = e^{z^2/4}\sqrt{\pi/2}\,\mathrm{erfc}(z/\sqrt{2}), \tag{3.50}$$

we find from Eq. (3.44)

$$E_\tau(q) = \tau\Delta + \tau\sqrt{\frac{2}{\pi}}\frac{\hbar v}{\ell}\frac{e^{-q^2/2}}{\mathrm{erfc}(-q/\sqrt{2})}, \qquad \tau = \pm 1, \tag{3.51}$$

where erfc(z) is the complementary error function. Since it is positive for any q, the zero-energy states emerge only for $\Delta < 0$. The analytical spectrum (3.51) is plotted, along with the corresponding numerical solution of Eq. (3.44), in Fig. 3.3.

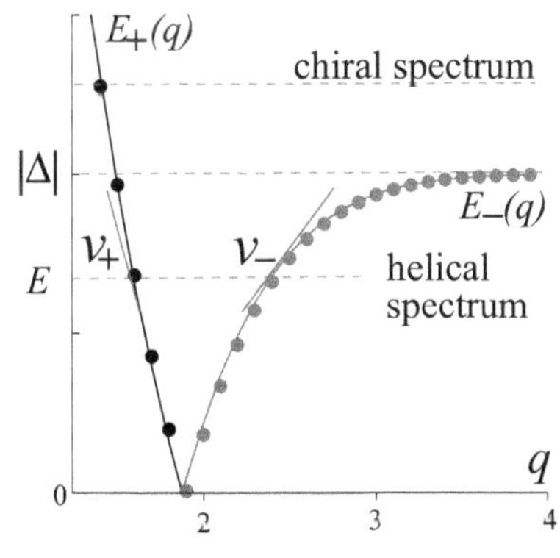

Figure 3.3 Edge-state energies $E_+(q)$ and $E_-(q)$ as function of dimensionless wave vector, q, for a strong magnetic field with $\hbar v/\ell|\Delta| = 10$. Points indicate numerical solution of Eq. (3.44), whereas solid lines are the analytical result, Eq. (3.51). Note that $E_+(q)$ and $E_-(q)$ are nonlinear and have different group velocities v_+ and v_-. The helical spectrum turns chiral at the band gap energy $E = |\Delta|$.

The most striking feature of the high-field spectrum (3.51) is its *nonlinearity*. In the zero field [see Eq. (2.105) and Fig. 2.5], the edge-state energy is linear in wave vector, k, so that the counter-propagating states with $\tau = \pm 1$ have equal velocities, v. As shown in Fig. 3.3, in the strong-field regime the edge-state velocities are no longer equal. Moreover, at energy E close to the gap $|\Delta|$ the $\tau = -$ state merges with the bulk Landau level and has zero velocity $v_- \to 0$. Above the gap there is only one propagating edge state with $\tau = +$. Therefore, upon crossing the gap energy the spectrum changes from helical to chiral, and the 2DTI state transforms into a dissipationless quantum Hall state [67].

3.1.4 *Spatial Distribution of Edge States*

Apart from the spectral nonlinearity, the strong magnetic field modifies the spatial distribution of the helical edge states. In the zero field case [see Eqs. (2.106) and (2.107)], the maximum of the edge-state functions is exactly at the boundary $y = 0$. This is no longer the case in the strong-field regime. The complete information on the spatial dependence of the edge states is contained in the matrix Green function (3.6). Inserting Eqs. (3.7), (3.8), (3.42), and (3.45) into Eq. (3.6) and eliminating the derivatives with the help of the recurrence relations for $U(a, z)$ [115], we find the full Green function matrix

$$\hat{G}_{\tau k}(z, z') = \begin{bmatrix} G_{\uparrow\uparrow,\tau k}(z, z') & G_{\uparrow\downarrow,\tau k}(z, z') \\ G_{\downarrow\uparrow,\tau k}(z, z') & G_{\downarrow\downarrow,\tau}(z, z') \end{bmatrix} \tag{3.52}$$

$$= \frac{\alpha(z, z') \begin{pmatrix} 1 & \beta(z') \\ \beta(z) & \beta(z)\beta(z') \end{pmatrix}}{E - \Delta\tau - (\hbar v/\ell)U(a, -q)/U(a+1, -q)}, \tag{3.53}$$

where functions $\alpha(z, z')$ and $\beta(z)$ are given by

$$\alpha(z, z') = \frac{U(a, z-q)U(a, z'-q)}{\ell U(a, -q)U(a+1, -q)}, \tag{3.54}$$

$$\beta(z) = \frac{U(a, -q)U(a+1, z-q)}{U(a+1, -q)U(a, z-q)}, \tag{3.55}$$

and parameters a and q are defined in Eqs. (3.13) and (3.18).

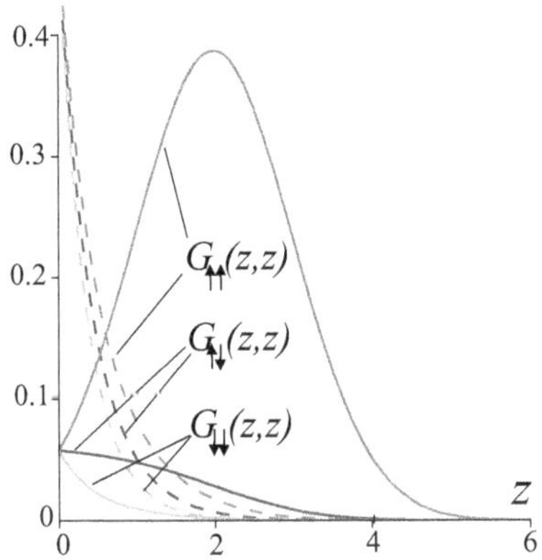

Figure 3.4 Typical spatial distributions of edge states in strong-field (solid curves) and weak-field (dashed curves) regimes defined by Eqs. (3.48) and (3.47), respectively; $E = 0$ and for $\Delta < 0$.

Figure 3.4 shows typical spatial dependences of the functions $G_{\uparrow\uparrow,\tau k}(z, z) \propto \alpha(z, z)$, $G_{\uparrow\downarrow,\tau k}(z, z) \propto \alpha(z, z)\beta(z)$ and $G_{\downarrow\downarrow,\tau k}(z, z) \propto \alpha(z, z)\beta^2(z)$ at zero energy $E = 0$. Interestingly, the $\uparrow$ component is no longer localized at the edge of the system ($z = 0$). Instead, it behaves almost like the lowest-Landau-level bulk wave function with the center at $z \sim q$. The other functions are small at $z \sim q$. Using Eqs. (3.49) and (3.50) and, consequently, $\beta \ll 1$, we obtain the strong-field asymptotic of Eq. (3.53) as

$$\hat{G}_{\tau k}(z, z') \approx \frac{\sigma_0 + \sigma_z}{2} G_{\tau,k}(z, z'), \qquad G_{\tau,k}(z, z') \approx \frac{\alpha(z, z')}{E - E_\tau(q)}, \tag{3.56}$$

where the edge-state spectrum $E_\tau(q)$ is given by Eq. (3.51), and function $\alpha(z, z')$ [see Eq. (3.54)] reduces to

$$\alpha(z, z') \approx \sqrt{\frac{2}{\pi}} \frac{e^{-(z-q)^2/4-(z'-q)^2/4}}{\ell \operatorname{erfc}(-q/\sqrt{2})}. \tag{3.57}$$

3.2 Backscattering, Conductance, and Topological Transitions

Since the helical edge states survive in a strong magnetic field, the results of Section 2.2.5 for ballistic transport hold also for the finite

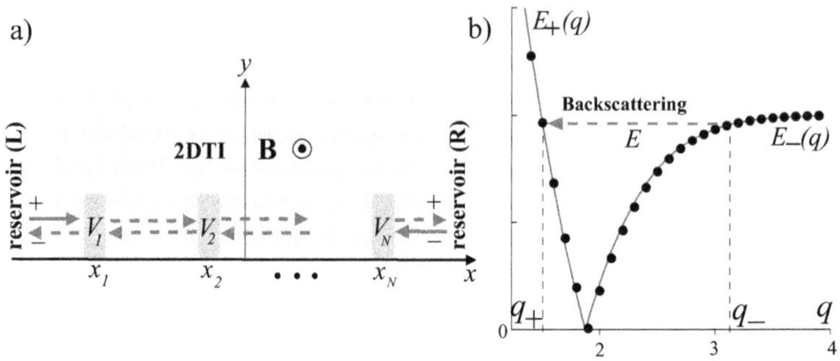

Figure 3.5 (a) 2DTI coupled to two carrier reservoirs and containing backscattering centers in the interior (shaded regions). (b) Illustration of a backscattering process between edge states with wave-vectors $q_\pm$ at energy E.

magnetic fields. For a single ballistic edge coupled to two carrier reservoirs, the longitudinal electric conductance is simply

$$g = \frac{e^2}{h}. \tag{3.58}$$

In the following section we discuss new phenomena that occur in the presence of impurities generating edge *backscattering*, whereby one edge state is scattered into the other one with the opposite chirality, as schematically depicted in Fig. 3.5. Such impurities must violate the TRS. In order to construct a theoretical model of the backscattering we first need to introduce the free edge-state propagator in real space.

3.2.1 *Free Edge-State Propagator*

Since in strong fields the edge-state width shrinks proportionally to the magnetic length ℓ, the free edge-state propagator can be described by the one-dimensional formula,

$$G_\tau(x, x') = \int \frac{dk}{2\pi} e^{ik(x-x')} \ell \int_0^\infty dz\, G_{\tau k}(z, z), \tag{3.59}$$

where $G_{\tau k}(z, z)$ is given by Eqs. (3.56) and (3.57). Taking into account that $\ell \int_0^\infty dz\, \alpha(z, z) = 1$ [115], we have from Eq. (3.59)

$$G_\tau(x, x') = \int \frac{dk}{2\pi} \frac{e^{ik(x-x')}}{E - E_\tau(q)}, \tag{3.60}$$

where k and q are related by Eq. (3.18), with eB chosen to be positive. The main contribution to the k-integral is expected to come from the vicinity of the point, $q_\tau(E, B)$, at which energy E crosses dispersion $E_\tau(q)$ (see also Fig. 3.5b),

$$E = E_\tau(q) \quad \rightarrow \quad q_\tau(E, B), \quad \tau = \pm. \tag{3.61}$$

We, therefore, linearize the edge dispersion [Eq. (3.51)] near fixed energy E,

$$E_\tau(q) \approx E - \frac{\hbar v_\tau}{2\ell}(q - q_\tau), \tag{3.62}$$

where velocity $v_\tau(E, B)$ is given by

$$v_\tau(E, B) = \frac{2\ell}{\hbar}(\tau|\Delta| + E)\, q_\tau(E, B). \tag{3.63}$$

This formula explicitly shows that $v_-(E, B) \rightarrow 0$ as $E \rightarrow |\Delta|$, in agreement with Fig. 3.5b. Using the linearized dispersion (3.62) in Eq. (3.60) we finally find

$$G_\tau(x, x') = \Theta([x - x']\tau)\frac{e^{ik_\tau(E,B)(x-x')}}{i\hbar|v_\tau(E, B)|}, \quad k_\tau(E, B) = -\frac{q_\tau(E, B)}{2\ell}. \tag{3.64}$$

This equation describes a plane wave propagating with wave-vector $k_\tau(\epsilon, B)$ and velocity $v_\tau(\epsilon, B)$. The nonzero values of the function $\Theta([x - x']\tau)$ determine the propagation direction, that is, edge-state chirality. For $\tau = +1$ the coordinate x lies on the right from x', that is, the state propagates from left to right (*right-mover*), whereas for $\tau = -1$ the coordinate x lies on the left from x', which corresponds to the *left-mover* propagating from right to left.

3.2.2 *Edge Scattering Matrix and Fisher–Lee Relation*

Now, we consider a disordered system connected to two carrier reservoirs on the left (L) and on the right (R), as shown in Fig. 3.5a. We introduce notations L^+ and R^- for the incoming (from the reservoirs) states, and L^- and R^+ for the states scattered toward the reservoirs (outgoing states). Transport characteristics of the disordered system can be inferred from the scattering matrix, $\hat{S}$, connecting the incoming and outgoing states,

$$\begin{bmatrix} L^- \\ R^+ \end{bmatrix}_{out} = \hat{S}\begin{bmatrix} L^+ \\ R^- \end{bmatrix}_{in}, \quad \hat{S} = \begin{bmatrix} r^{-+}_{LL} & t^{--}_{LR} \\ t^{++}_{RL} & r^{+-}_{RR} \end{bmatrix}, \tag{3.65}$$

where $r^{-+(+-)}_{LL(RR)}$ and $t^{++(--)}_{RL(LR)}$ are the all possible reflection and transmission amplitudes. Without loosing generality, we assume that the current is injected into the right-moving (+) channel incoming from the L reservoir. Then, the electric conductance is proportional to the transmission probability, $|t^{++}_{RL}|^2$, from L to R

$$g = \frac{e^2}{h}\left|t^{++}_{RL}\right|^2. \tag{3.66}$$

where e^2/h is the conductance of a perfectly transparent channel. The transmission amplitude can be calculated from the Green function of the disordered system using Fisher–Lee relation [116],

$$t^{++}_{RL} = i\hbar|v_+|\mathcal{G}_{++}(x \in R, x' \in L), \tag{3.67}$$

where the Green function $\mathcal{G}_{++}(x, x')$ is the diagonal element of the Green function matrix

$$\hat{\mathcal{G}}(x, x') = \begin{bmatrix} \mathcal{G}_{++}(x, x') & \mathcal{G}_{+-}(x, x') \\ \mathcal{G}_{-+}(x, x') & \mathcal{G}_{--}(x, x') \end{bmatrix}. \tag{3.68}$$

It is a matrix in space of the right (+)- and left (−) -moving states. Its off-diagonal elements $\mathcal{G}_{+-}(x, x')$ and $\mathcal{G}_{-+}(x, x')$ originate from edge-state backscattering. We model the backscatterers by the sum of N matrix potentials localized at positions x_n ($n = 1..N$):

$$\hat{V}(x) = \sum\nolimits_{n=1..N} \begin{bmatrix} 0 & V_n\delta(x - x_n) \\ V_n\delta(x - x_n) & 0 \end{bmatrix}. \tag{3.69}$$

It is off-diagonal in right-left mover space, leading to interchannel scattering with the strength quantified by matrix element V_n. We note that potential $\hat{V}(x)$ does not commute with the TRS operation.

3.2.3 *Solution of Dyson Equation. Conductance*

In the presence of potential $\hat{V}(x)$ (3.69) the edge-state propagator obeys the equation

$$\hat{\mathcal{G}}(x, x') = \hat{G}(x, x') + \sum_{n=1..N} \hat{G}(x, x_n) \begin{bmatrix} 0 & V_n \\ V_n & 0 \end{bmatrix} \hat{\mathcal{G}}(x_n, x'), \tag{3.70}$$

where

$$\hat{G}(x, x') = \begin{bmatrix} G_+(x, x') & 0 \\ 0 & G_-(x, x') \end{bmatrix} \tag{3.71}$$

is the Green function matrix in the absence of scattering [see Eq. (3.64)]. From Eqs. (3.70) and (3.71) we have two coupled equations for the diagonal, $\mathcal{G}_{++}(x, x')$ and off-diagonal, $\mathcal{G}_{-+}(x, x')$, elements:

$$\mathcal{G}_{++}(x, x') = G_{+}(x, x') + \sum_{n} G_{+}(x, x_n) V_n \mathcal{G}_{-+}(x_n, x'), \qquad (3.72)$$

$$\mathcal{G}_{-+}(x, x') = \sum_{n} G_{-}(x, x_n) V_n \mathcal{G}_{++}(x_n, x'). \qquad (3.73)$$

Eliminating $\mathcal{G}_{-+}(x, x')$ yields a closed Dyson equation for $\mathcal{G}_{++}(x, x')$:

$$\mathcal{G}_{++}(x, x') = G_{+}(x, x') + \sum\nolimits_{n,m} G_{+}(x, x_n) V_n G_{-}(x_n, x_m) V_m \mathcal{G}_{++}(x_m, x'). \qquad (3.74)$$

With known function $G_\tau(x, x')$ [see Eq. (3.64)] and for not large N, we can solve Eq. (3.74) analytically and calculate g. Leaving aside this algebra, we present the final results for the particular cases $N = 1$, 2 and 3:

$$g = \frac{e^2}{h}\left(1 + \frac{V_1^2}{\hbar^2 |v_+ v_-|}\right)^{-2}, \qquad (3.75)$$

$$g = \frac{e^2}{h}\left|1 + \frac{V_1^2 V_2^2}{\hbar^4 v_+^2 v_-^2} + \frac{V_1^2 + V_2^2 + V_1 V_2 e^{iQd_{12}}}{\hbar^2 |v_+ v_-|}\right|^{-2}, \qquad (3.76)$$

$$g = \frac{e^2}{h}\left|1 + \frac{V_1^2 V_2^2 V_3^2}{\hbar^6 |v_+ v_-|^3} + \frac{\sum_{n=1}^{3} V_n^2 + V_1 V_2 e^{iQd_{12}} + V_1 V_3 e^{iQd_{13}} + V_2 V_3 e^{iQd_{23}}}{\hbar^2 |v_+ v_-|} + \frac{V_1^2 V_2^2 + V_1^2 V_3^2 + V_2^2 V_3^2 + V_1^2 V_2 V_3 e^{iQd_{23}}}{\hbar^4 v_+^2 v_-^2}\right|^{-2}, \qquad (3.77)$$

where $Q = k_+ - k_-$ and $d_{nm} = x_m - x_n$. The wave-vectors $k_\pm$ [see Eq. (3.64)] and velocities $v_\pm$ [see Eq. (3.63)] are obtained by solving numerically Eq. (3.61).

3.2.4 *Transition from 2DTI to a Quantum Hall State*

The interesting result is that all V-terms in Eqs. (3.75)–(3.77) diverge when dispersion $E_-(q)$ becomes flat at the gap energy with vanishing velocity, $v_- \to 0$ (see Fig. 3.5b). The most divergent terms are, clearly, those with the highest power of Vs. Generalizing this observation for arbitrary N, we can write the most divergent term as

$$\frac{V_1^2 \times \ldots \times V_N^2}{\hbar^{2N}|v_+ v_-|^N} \tag{3.78}$$

It arises from simultaneous scattering from N potentials. The divergence of Eq. (3.78) means that the edge backscattering is greatly enhanced, leading to the suppressed conductance

$$g(E, B) \approx \frac{e^2}{h} \times \frac{\hbar^{4N}|v_+ v_-|^{2N}}{(V_1 \times \ldots \times V_N)^4} \ll \frac{e^2}{h}, \quad v_- \to 0. \tag{3.79}$$

Using Eq. (3.63) for the velocities $v_\pm(E, B)$ we find the qualitative energy and field dependence of the conductance near the band gap:

$$g(E, B) \propto \frac{(|\Delta| - |E|)^{2N}}{B^{2N}}, \quad |E| \to |\Delta|. \tag{3.80}$$

It shows a power-law magnetic field dependence B^{-2N}, determined by the number N of backscattering centers on the edge. This suggests a simple way to detect such defects in quasi-ballistic 2DTIs using standard magnetoresistance measurements.

In a wider range of energies and fields, the typical behavior of the conductance can be understood from Eq. (3.76) for two backscattering centers on the edge. Figure 3.6 shows the energy dependence of g at fixed magnetic field $B = 0.5$ T. Independently of disorder strength, the conductance drops to zero as the gap energy $|\Delta| = 1.5$ meV is approached. For relatively weak $V_{1,2}$ (upper curve) the transition is sharp because the initial conductance is close to the ballistic value e^2/h. An increase in disorder strength $V_{1,2}$ (lower curve) smears the transition.

The exact vanishing of the longitudinal (dissipative) conductance indicates a *topological transition* from the 2DTI to a chiral quantum Hall state. The longitudinal conductance must vanish at $E = |\Delta|$

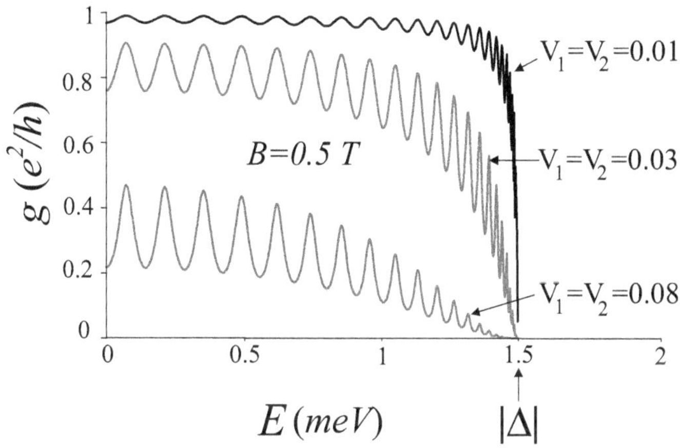

Figure 3.6 Longitudinal conductance [see Eq. (3.76)] versus energy for $B = 0.5$ T, $|\Delta| = 1.5$ meV and $d_{12} = 3$ μm. Disorder potential strengths $V_{1,2}$ are in units of meV·μm. The exact vanishing of g indicates a topological transition from the 2DTI to a dissipationless quantum Hall state (adapted from Ref. [67]).

since higher energies $E > |\Delta|$ fall into a Landau gap in which the quantum Hall state is dissipationless and has only nonzero Hall conductivity. The transition is accompanied by Fabry–Perot-type oscillations due to interference of the counter-propagating channels which acquire the energy- and field-dependent phase difference $(k_+ - k_-)d_{12}$, in scattering between the defects. Summarizing our findings, we emphasize two chief factors needed for observing the topological transition studied above: edge backscattering and strong quantizing magnetic fields.

3.3 Problems

(1) Prove the commutation relation

$$P_x P_y - P_y P_x = \frac{ie\hbar}{c} B \tag{3.81}$$

for the components of the momentum operator $\boldsymbol{P}$ in the presence of a magnetic field [see Eq. (3.3)].

(2) Use Eq. (3.81) and the plane-wave expansion in Eq. (3.11) to derive the ordinary differential equations

$$\left[\partial_y^2 - \frac{\Delta^2 - E^2}{(\hbar v)^2} + \frac{eB}{c\hbar} - \left(k + \frac{eBy}{c\hbar}\right)^2\right] G_{\uparrow\uparrow,\tau} = \frac{E + \Delta\tau}{(\hbar v)^2}\delta(y - y'), \tag{3.82}$$

and

$$\left[\partial_y^2 - \frac{\Delta^2 - E^2}{(\hbar v)^2} - \frac{eB}{c\hbar} - \left(k + \frac{eBy}{c\hbar}\right)^2\right] G_{\downarrow\downarrow,\tau} = \frac{E - \Delta\tau}{(\hbar v)^2}\delta(y - y'). \tag{3.83}$$

Show that for the dimensionless coordinate $z = y/\ell$ and wave vector $q = -2k\ell\text{sgn}(eB)$ [see Eqs. (3.16) and (3.18)], the equations above transform to Eqs. (3.12) and (3.14).

(3) Consider the second order differential equation

$$[\partial_x^2 + u(x)] f(x) = 0, \tag{3.84}$$

with two linearly independent solutions $f_1(x)$ and $f_2(x)$. Prove that for any $u(x)$ the Wronskian

$$W = \partial_x f_1(x) \cdot f_2(x) - f_1(x) \cdot \partial_x f_2(x) \tag{3.85}$$

is independent of x. Let us now introduce the function of two variables

$$G(x, x') = \frac{1}{W}\left[\Theta(x - x') f_1(x) f_2(x') + \Theta(x' - x) f_1(x') f_2(x)\right]. \tag{3.86}$$

Prove that $G(x, x')$ satisfies the equation

$$[\partial_x^2 + u(x)] G(x, x') = \delta(x - x'). \tag{3.87}$$

Equation (3.86) is, therefore, the Green function of a general second order differential equation.

Chapter 4

Three-Dimensional Topological Insulators

The helical edge states of 2DTIs have higher-dimensional analogues, Dirac-like 2D states at a surface of a bulk material. A bulk system with the Dirac surface states inside the band gap is a three-dimensional topological insulator (3DTI). To elucidate the origin of the surface states, in this section we introduce a microscopic model based on the Kane $\boldsymbol{k} \cdot \boldsymbol{p}$ Hamiltonian with a nontrivial boundary condition. We shall see that out of all bulk degrees of freedom the surface state inherits only the electron spin. Similar to the helical edge states, the spin of the surface carrier is locked to the direction of its momentum. In the 2D surface geometry the spin–momentum locking gives rise to a new transport phenomenon, weak antilocalization. We shall discuss both a simple qualitative picture and the detailed theory of this phenomenon.

Topological Insulators: The Physics of Spin Helicity in Quantum Transport
Grigory Tkachov

ISBN 978-981-4613-25-5 (Hardcover), 978-981-4613-26-2 (eBook)
www.panstanford.com

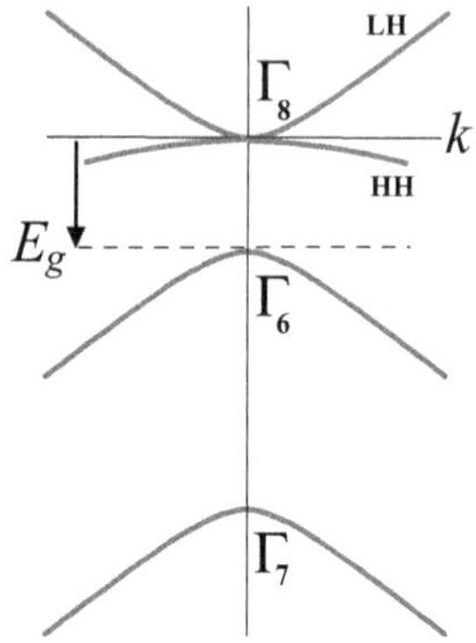

Figure 4.1 Schematic band structure of bulk unstrained HgTe.

4.1 Helical Surface States. Strained HgTe as a 3DTI

4.1.1 *Kane Hamiltonian and Boundary Conditions*

To be concrete, we consider bulk HgTe. It is a textbook example of a narrow-gap semiconductor with a multiband electronic structure. As already mentioned in Chapter 2, the conduction and valence bands of HgTe are located close to the center (the $k = 0$ point) of the Brillouin zone and have the s and p symmetries deriving from the atomic orbitals of Hg and Te, respectively. The $\boldsymbol{k} \cdot \boldsymbol{p}$ and spin–orbit splitting result in the three bands at $k = 0$: an s-type double degenerate Γ_6, a p-type four-fold degenerate Γ_8 with $j = 3/2$, and a p-type double degenerate Γ_7 with $j = 1/2$ (Ref. [53] and Fig. 4.1). Away from the zone center, the Γ_8 band splits into the light hole (LH) and heavy hole (HH) bands with the total angular momentum projections $\pm 1/2$ and $\pm 3/2$, respectively.

The degeneracy of the LH and HH at the Fermi point $k = 0$ is lifted by strain typically present in HgTe grown on a substrate with a different lattice constant [33]. Transport [33, 119, 120] and magneto-optical [117, 118] experiments indicate that the electronic properties of strained HgTe are determined mainly by the topological surface states. We intend to derive these states from the Kane $\boldsymbol{k} \cdot \boldsymbol{p}$ Hamiltonian (Ref. [121]). For our study, the coupling to the remote Γ_7 band is not essential, so that one can work with the six-component envelope wave function

$$f(\boldsymbol{r}) = \begin{bmatrix} f_1(\boldsymbol{r}) \\ f_2(\boldsymbol{r}) \\ f_3(\boldsymbol{r}) \\ f_4(\boldsymbol{r}) \\ f_5(\boldsymbol{r}) \\ f_6(\boldsymbol{r}) \end{bmatrix} \equiv \begin{bmatrix} |\Gamma_6, +\frac{1}{2}\rangle \\ |\Gamma_6, -\frac{1}{2}\rangle \\ |\Gamma_8, +\frac{3}{2}\rangle \\ |\Gamma_8, +\frac{1}{2}\rangle \\ |\Gamma_8, -\frac{1}{2}\rangle \\ |\Gamma_8, -\frac{3}{2}\rangle \end{bmatrix}. \qquad (4.1)$$

Here f_1 and f_2 correspond to the s-type states of Γ_6 with the spin projections $\pm 1/2$, f_3, and f_6 represent the HH, while f_4 and f_5 - the LH states of Γ_8. The corresponding 6×6 Kane Hamiltonian for the $\boldsymbol{k} \cdot \boldsymbol{p}$ - coupled Γ_6 and Γ_8 bands can be written as

$$\hat{\mathcal{H}} = \begin{bmatrix} E_g & 0 & -P_+ & P_z & P_-/\sqrt{3} & 0 \\ 0 & E_g & 0 & -P_+/\sqrt{3} & P_z & P_- \\ -P_- & 0 & U & 0 & 0 & 0 \\ P_z & -P_-/\sqrt{3} & 0 & V & 0 & 0 \\ P_+/\sqrt{3} & P_z & 0 & 0 & V & 0 \\ 0 & P_+ & 0 & 0 & 0 & U \end{bmatrix}. \qquad (4.2)$$

Here, E_g is the direct gap between the Γ_6 and Γ_8 bands. It is negative, $E_g = -300$ meV, since HgTe has the inverted band structure [53]. The matrix elements $P_\pm$ and P_z are linear functions of the components of the electron momentum,

$$P_\pm = \sqrt{\frac{1}{2}} P\,(-i\partial_x \pm \partial_y), \qquad P_z = -\sqrt{\frac{2}{3}} P\, i\partial_z, \qquad (4.3)$$

where P is the Kane parameter. The diagonal entries U and V are the energies of the HH and LH bands, respectively, measured from the Fermi level at $E = 0$. The case $U \neq V$ corresponds to the strained HgTe, with an energy gap $V - U$ between the LH and HH states. The terms of higher order in the electron momentum are omitted in Eq. (4.2) because near $k = 0$ they produce small corrections to the linear surface dispersion.

We assume that the material occupies the half-space $z < 0$ with the boundary to the vacuum at $z = 0$. The function $f(\boldsymbol{r})$ must satisfy a boundary condition at $z = 0$ that ensures the vanishing of the normal component of the particle current density

$$j_z(\boldsymbol{r}) = \tilde{f}(\boldsymbol{r})\hat{v}_z f(\boldsymbol{r}), \qquad \hat{v}_z = (z\hat{\mathcal{H}} - \hat{\mathcal{H}}z)/i\hbar, \qquad (4.4)$$

where $\tilde{f}(\mathbf{r}) = [f_1^*(\mathbf{r}), f_2^*(\mathbf{r}), f_3^*(\mathbf{r}), f_4^*(\mathbf{r}), f_5^*(\mathbf{r}), f_6^*(\mathbf{r})]$ is the conjugated wave function, and the normal component of the velocity, $\hat{v}_z$, is given for Hamiltonian (4.2) by

$$\hat{v}_z = \sqrt{\frac{2}{3}}\frac{P}{\hbar}\begin{bmatrix} 0&0&0&1&0&0 \\ 0&0&0&0&1&0 \\ 0&0&0&0&0&0 \\ 1&0&0&0&0&0 \\ 0&1&0&0&0&0 \\ 0&0&0&0&0&0 \end{bmatrix}. \tag{4.5}$$

The equation above shows that the current j_z is carried by the coupled s and LH states. The HH drops out of Eq. (4.4). Therefore, the boundary problem can be reduced to the s - LH sector. In order to exclude the HH states, we write the eigenvalue equation $\hat{\mathcal{H}} f(\mathbf{r}) = E f(\mathbf{r})$ in components

$$E_g f_1 - P_+ f_3 + P_z f_4 + (P_-/\sqrt{3}) f_5 = E f_1, \tag{4.6}$$
$$E_g f_2 - (P_+/\sqrt{3}) f_4 + P_z f_5 + P_- f_6 = E f_2, \tag{4.7}$$
$$-P_- f_1 + U f_3 = E f_3, \tag{4.8}$$
$$P_z f_1 - (P_-/\sqrt{3}) f_2 + V f_4 = E f_4, \tag{4.9}$$
$$(P_+/\sqrt{3}) f_1 + P_z f_2 + V f_5 = E f_5, \tag{4.10}$$
$$P_+ f_2 + U f_6 = E f_6. \tag{4.11}$$

From Eqs. (4.8) and (4.11) we find the HH components as

$$f_3 = \frac{P_-}{U - E} f_1, \qquad f_6 = -\frac{P_+}{U - E} f_2. \tag{4.12}$$

Inserting these into the left-hand side of Eqs. (4.6), (4.7), (4.9), and (4.10), we obtain a closed system for the s and LH components, which in the compact notations reads

$$\hat{H}\psi(\mathbf{r}) = E\psi(\mathbf{r}), \tag{4.13}$$

where $\psi(\mathbf{r})$ is the four-component wave function

$$\psi(\mathbf{r}) = \begin{bmatrix} f_1(\mathbf{r}) \\ f_2(\mathbf{r}) \\ f_4(\mathbf{r}) \\ f_5(\mathbf{r}) \end{bmatrix}, \tag{4.14}$$

and $\hat{H}$ is the effective Hamiltonian for the coupled s and LH states

$$\hat{H} = \begin{bmatrix} E_g - \frac{P_+P_-}{U-E} & 0 & P_z & \frac{P_-}{\sqrt{3}} \\ 0 & E_g - \frac{P_-P_+}{U-E} & \frac{-P_+}{\sqrt{3}} & P_z \\ P_z & \frac{-P_-}{\sqrt{3}} & V & 0 \\ \frac{P_+}{\sqrt{3}} & P_z & 0 & V \end{bmatrix}. \tag{4.15}$$

We see that the coupling to the HH leads to the second order corrections in the electron momentum. These terms can be neglected in the vicinity of the $k = 0$ point.

In the following, we adopt the linearized version of Hamiltonian (4.15) in the Pauli matrix representation

$$\hat{H} \approx \frac{E_g + V}{2}\sigma_0\tau_0 + \frac{E_g - V}{2}\sigma_0\tau_z - i\mathcal{B}\sigma_0\tau_x\partial_z + i\mathcal{A}(\sigma_y\partial_x - \sigma_x\partial_y)\tau_y, \tag{4.16}$$

where the matrices σ_j and τ_j act in spin and s - LH spaces, respectively, and the constants $\mathcal{A}$ and $\mathcal{B}$ are defined as

$$\mathcal{A} = \sqrt{\frac{1}{6}}P, \qquad \mathcal{B} = \sqrt{\frac{2}{3}}P. \tag{4.17}$$

The normal component of the current density [Eqs. (4.4) and (4.5)] in the basis of Eq. (4.14) is

$$j_z(\boldsymbol{r}) = \frac{\mathcal{B}}{\hbar}\tilde{\psi}(\boldsymbol{r})\sigma_0\tau_x\psi(\boldsymbol{r}). \tag{4.18}$$

The matrix structure of the equation above suggests the following boundary condition at the surface $z = 0$:

$$\psi(x, y, 0) = \sigma_0\tau_y\psi(x, y, 0). \tag{4.19}$$

It is easy to see that the current indeed vanishes at the boundary:

$$j_z(x, y, 0) = \frac{\mathcal{B}}{\hbar}\tilde{\psi}(x, y, 0)\sigma_0\tau_x\psi(x, y, 0) \tag{4.20}$$

$$= \frac{\mathcal{B}}{\hbar}\tilde{\psi}(x, y, 0)\sigma_0\tau_y\,\sigma_0\tau_x\,\sigma_0\tau_y\psi(x, y, 0) \tag{4.21}$$

$$= -\frac{\mathcal{B}}{\hbar}\tilde{\psi}(x, y, 0)\sigma_0\tau_x\psi(x, y, 0) = -j_z(x, y, 0), \tag{4.22}$$

which necessarily leads to $j_z(x, y, 0) = 0$. Drawing a parallel with the edge states considered in Chapter 2, one can prove that the boundary condition (4.19) is equivalent to a topological gap domain wall at $z = 0$.

4.1.2 *Surface-State Dispersion, Helicity, and Decay Length*

The next step is to solve the boundary problem described by Eqs. (4.13), (4.16), and (4.19). We are looking for the solution propagating parallel to the surface $z = 0$ and decaying exponentially for $z \to -\infty$:

$$\psi(x, y, z) = \Phi_{\boldsymbol{k}} e^{i(k_x x + k_y y) + z/\lambda}, \tag{4.23}$$

$$\Phi_{\boldsymbol{k}} = \sigma_0 \tau_y \Phi_{\boldsymbol{k}}. \tag{4.24}$$

Again, the constraint (4.24) on the spinor $\Phi_{\boldsymbol{k}}$ follows from the boundary condition (4.19). Inserting Eq. (4.23) into Eq. (4.13) and using Eq. (4.24), we have

$$\left[\frac{E_g + V}{2} + \mathcal{A}(\sigma_x k_y - \sigma_y k_x)\right] \Phi_{\boldsymbol{k}} + \left[\frac{E_g - V}{2} + \frac{\mathcal{B}}{\lambda}\right] \tau_z \Phi_{\boldsymbol{k}} = E\Phi_{\boldsymbol{k}}. \tag{4.25}$$

The decay length is determined from the requirement for the second term to vanish identically,

$$\lambda = \frac{2\mathcal{B}}{V - E_g}. \tag{4.26}$$

Then, Eq. (4.25) reduces to

$$\hat{H}_{surf}\Phi_{\boldsymbol{k}} = E\Phi_{\boldsymbol{k}}, \qquad \hat{H}_{surf} = \frac{E_g + V}{2} + \mathcal{A}(\sigma_x k_y - \sigma_y k_x), \tag{4.27}$$

where $\hat{H}_{surf}$ can be identified as the surface Hamiltonian. Its eigenstates coincide with those of the helicity matrix $(\boldsymbol{\sigma} \times \boldsymbol{n})_z$,

$$(\boldsymbol{\sigma} \times \boldsymbol{n})_z \Phi_{\boldsymbol{k}s} = s\Phi_{\boldsymbol{k}s}, \qquad \boldsymbol{n} = \frac{\boldsymbol{k}}{|\boldsymbol{k}|}, \qquad s = \pm 1, \tag{4.28}$$

while the energy eigenvalues are

$$E_{\boldsymbol{k}s} = \frac{E_g + V}{2} + s\mathcal{A}|\boldsymbol{k}|. \tag{4.29}$$

Let us summarize the results of this section:

- The surface state occurs in the middle of the gap between the bulk Γ_6 (s-type) and LH (p-type) bands and has a Dirac-like dispersion. The 3DTI phase is realized in the energy gap $V - U$ between the LH and HH bands, where only the surface state exists (see Fig. 4.2). Although our linear model is justified only in the vicinity of the $k = 0$ point, the conclusion regarding the 3DTI phase is still correct and agrees with numerical calculations based on extended models including higher order k-terms [33, 186].

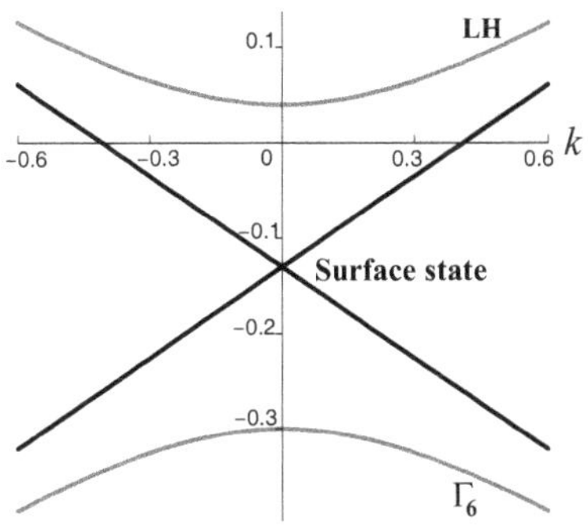

Figure 4.2 Surface and bulk bands of strained HgTe versus in-plane wave vector k (in nm^{-1}) from the reduced Hamiltonian (4.16) with $P = 780$ meV·nm, $E_g = -300$ meV, $V = 40$ meV, and $U = 0$ meV. Shaded area indicates the strain-induced gap in which the 3DTI state is realized.

- In weakly strained HgTe with $V \ll |E_g|$, the surface-state decay length is determined by the gap energy E_g and the Kane matrix element P [see Eqs. (4.17) and (4.26)]. For the typical values $E_g = -300$ meV and $P = 780$ meV·nm [33, 53] the decay length is of order of a few nanometers

$$\lambda \approx 4 \text{ nm}. \tag{4.30}$$

- Because of the topological constraint (4.24) imposed by the boundary condition, the surface state inherits only a half of the bulk degrees of freedom, which is the electron spin in Eqs. (4.27) and (4.28). The spin helicity has major implications for electron transport in TI materials.

4.2 Bi-Based 3DTIs and thin TI Materials

4.2.1 *Surface States in Bi_2Se_3 and Bi_2Te_3*

Another large group of materials that host topological surface states is Bi-based compounds, such as $Bi_{1-x}Sb_x$, Bi_2Se_3, and Bi_2Te_3 [22, 28, 29]. Both band structure calculations [22] and ARPES measurements [28, 29] reveal Dirac-like states at the surface of these materials. The low-energy dispersion and other properties of the surface carriers can be derived from the Hamiltonian of a bulk material [22] in the same way as done in Section 4.1 for HgTe. An illustrative example is the surface state in $Bi_2Se(Te)_3$ described by the Hamiltonian [36, 37]

$$\hat{H}^s_{\boldsymbol{k}} = \mathcal{A}(\sigma_x k_y - \sigma_y k_x) + \mathcal{M}_{\boldsymbol{k}}\sigma_z + \mathcal{D}\boldsymbol{k}^2 - E_F, \tag{4.31}$$

$$\mathcal{M}_{\boldsymbol{k}} = \frac{W}{2}(k_+^3 + k_-^3), \qquad k_\pm = k_x \pm i k_y, \tag{4.32}$$

where $\mathcal{A} > 0$, $\mathcal{D} > 0$, and $W > 0$ are band structure parameters and E_F is the Fermi energy. Equation (4.31) is an extended version of Hamiltonian (4.16), where the quadratic term accounts for the broken particle-hole symmetry, whereas the cubic one $\mathcal{M}_{\boldsymbol{k}}$ reflects hexagonal warping of the surface bands in $Bi_2Se(Te)_3$ crystals [38, 39].

Upon unitary transformation

$$\hat{H}^s_{\boldsymbol{k}} \to U \hat{H}^s_{\boldsymbol{k}} U^\dagger, \qquad U = \begin{bmatrix} i & 0 \\ 0 & 1 \end{bmatrix}, \tag{4.33}$$

Hamiltonian (4.31) takes the form:

$$\hat{H}^s_{\boldsymbol{k}} = \mathcal{A}\boldsymbol{\sigma} \cdot \boldsymbol{k} + \mathcal{M}_{\boldsymbol{k}}\sigma_z + \mathcal{D}\boldsymbol{k}^2 - E_F. \tag{4.34}$$

The eigenvalues of Hamiltonian (4.34) yield the surface spectrum

$$E_\pm(k, \phi_{\boldsymbol{n}}) = \mathcal{D}k^2 - E_F \pm \sqrt{\mathcal{A}^2 k^2 + \frac{W^2 k^6}{2}(1 + \cos 6\phi_{\boldsymbol{n}})}. \tag{4.35}$$

It depends on the momentum direction specified by angle $\phi_{\boldsymbol{n}}$,

$$\boldsymbol{n} = \frac{\boldsymbol{k}}{|\boldsymbol{k}|} = (\cos\phi_{\boldsymbol{n}}, \sin\phi_{\boldsymbol{n}}, 0). \tag{4.36}$$

The surface dispersion relation (4.35) is plotted in Fig. 4.3.

4.2.2 *Thin TI Materials as 2D Helical Metals*

Another representative example of quasi-2D helical systems is thin TI materials in the metallic regime where the Fermi level crosses the bulk conduction (or valence) band. This can be realized in HgTe quantum wells (Refs. [54, 122]) and ultrathin films of $Bi_2Se(Te)_3$ (Refs. [40, 41]). Since the low-energy theories [4, 50, 51] for these materials share many similarities, the term "TI thin film" used below refers to both HgTe quantum wells and ultrathin films of $Bi_2Se(Te)_3$.

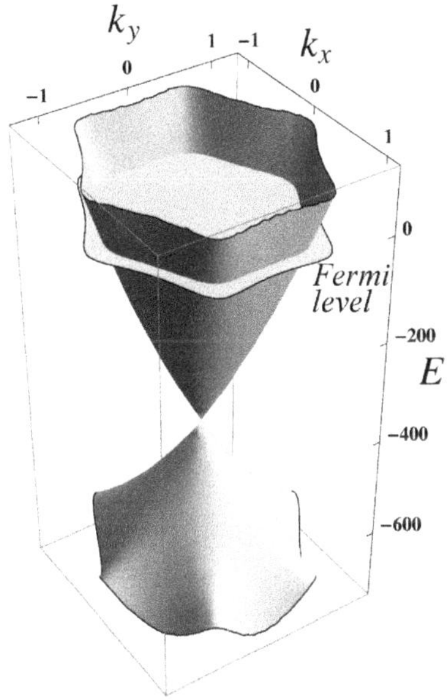

Figure 4.3 Energy bands (in meV) of a TI surface state [see Eq. (4.35)] versus in-plane wave-numbers, k_x and k_y (in nm^{-1}). Cross section at $E = 0$ corresponds to the Fermi level; $\mathcal{A} = 300$ meV·nm, $\mathcal{D} = 75$ meV·nm^2, $E_F = 400$ meV, and $W = 200$ meV·nm^3.

The low-energy Hamiltonian for a TI thin film, $\hat{H}^f_{\boldsymbol{k}}$, can be derived using the same procedure as introduced in Section 2.2 of Chapter 2 (see also Fig. 2.1). We start with the uncoupled top and bottom TI surfaces described by Hamiltonians $\pm\mathcal{A}\boldsymbol{\sigma}\cdot\boldsymbol{k}+\mathcal{D}\boldsymbol{k}^2 - E_F$. As the TI thickness is reduced, the two surfaces become coupled, which can be modeled by the Hamiltonian

$$\hat{H}^f_{\boldsymbol{k}} = \begin{bmatrix} \mathcal{A}\boldsymbol{\sigma}\cdot\boldsymbol{k}+\mathcal{D}\boldsymbol{k}^2 - E_F & \mathcal{M}_{\boldsymbol{k}}\sigma_0 \\ \mathcal{M}_{\boldsymbol{k}}\sigma_0 & -\mathcal{A}\boldsymbol{\sigma}\cdot\boldsymbol{k}+\mathcal{D}\boldsymbol{k}^2 - E_F \end{bmatrix}, \qquad (4.37)$$

where $\mathcal{M}_{\boldsymbol{k}}$ is the hybridization energy. The TRS requires $\mathcal{M}_{\boldsymbol{k}}$ to be an even function of $\boldsymbol{k}$, e.g.

$$\mathcal{M}_{\boldsymbol{k}} = \Delta + \mathcal{B}\boldsymbol{k}^2, \qquad (4.38)$$

where Δ and $\mathcal{B} > 0$ are band structure parameters. After the unitary transformation

$$\hat{H}^f_{\boldsymbol{k}} \to U\,\hat{H}^f_{\boldsymbol{k}}U^{\dagger}, \qquad \hat{U} = \frac{1}{\sqrt{2}}\begin{bmatrix} \sigma_0 & \sigma_z \\ \sigma_z & -\sigma_0 \end{bmatrix}, \qquad (4.39)$$

Hamiltonian (4.37) takes a block-diagonal form [4, 50, 51]

$$\hat{H}_{\boldsymbol{k}}^{f} = \begin{bmatrix} \hat{H}_{\boldsymbol{k}+} & 0 \\ 0 & \hat{H}_{\boldsymbol{k}-} \end{bmatrix}, \tag{4.40}$$

with $H_{\boldsymbol{k}\pm}$ given by

$$H_{\boldsymbol{k}\pm} = \pm\mathcal{A}\boldsymbol{\sigma}\cdot\boldsymbol{k} \pm \mathcal{M}_{\boldsymbol{k}}\sigma_z + \mathcal{D}\boldsymbol{k}^2 - E_F. \tag{4.41}$$

Unlike the surface model (4.34), here the term $\mathcal{M}_{\boldsymbol{k}}$ plays the role of the Dirac mass. The eigenvalues of Hamiltonians (4.41) are

$$E^{(1,2)}(k) = \mathcal{D}k^2 - E_F \pm \sqrt{\mathcal{A}^2k^2 + (\Delta + \mathcal{B}k^2)^2}, \tag{4.42}$$

$$E^{(3,4)}(k) = \mathcal{D}k^2 - E_F \mp \sqrt{\mathcal{A}^2k^2 + (\Delta + \mathcal{B}k^2)^2}, \tag{4.43}$$

where $E^{(1,4)}$ and $E^{(2,3)}$ form double-degenerate conduction and valence bands, respectively. The bands are separated by an energy gap Δ at $\boldsymbol{k} = 0$, as illustrated in Fig. 4.4.

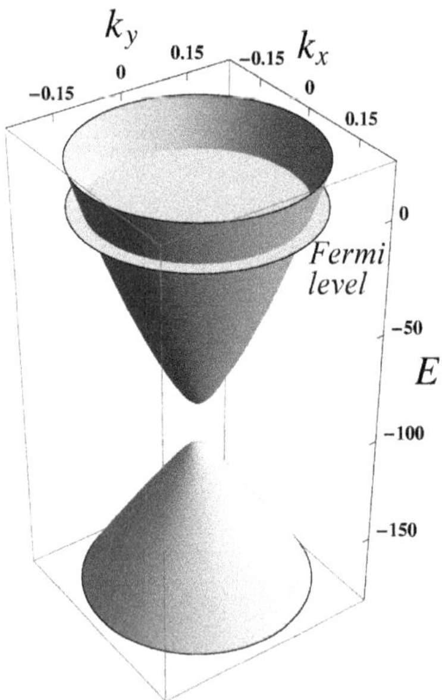

Figure 4.4 Energy bands (in meV) of a TI thin film [see Eqs. (4.42) and (4.43)] versus in-plane wave-numbers, k_x and k_y (in nm^{-1}). Cross section at $E = 0$ corresponds to the Fermi level; $\mathcal{A} = 380$ meV·nm, $\mathcal{B} = 850$ meV·nm^2, $\mathcal{D} = 350$ meV·nm^2, $\mathcal{M} = -10$ meV, and $E_F = 100$ meV.

The energy spectra for the TI surface (Fig. 4.3) and thin film (Fig. 4.4) are topologically distinct at the Dirac point $\boldsymbol{k} = 0$, which goes back to the different symmetries of the "mass" terms in Eqs. (4.32) and (4.38). This distinction manifests itself in quantum transport properties of these materials.

4.3 Electron Transport in Disordered TI Materials

The rest of this chapter deals with transport properties of 2D helical states in disordered TI materials. The main source of disorder in these materials is randomly distributed static ionized impurities generating an inhomogeneous spin-independent potential, $V(\boldsymbol{r})$, for mobile charge carriers. As in the 1D case, the TRS-invariant potential $V(\boldsymbol{r})$ cannot cause 180° backscattering between surface states with opposite wave-vectors, $\boldsymbol{k}$ and $-\boldsymbol{k}$. However, unlike the 1D case, scattering at any other angle is no longer forbidden, and therefore, through multiple collisions with impurities 2D helical carriers experience diffusion on a disordered surface (see also Fig. 4.5). The 2D diffusion of charged particles leads to a high electric conductivity, σ_c, compared with the conductance quantum,

$$\sigma_c \gg \frac{e^2}{h}. \tag{4.44}$$

The system of helical carriers in this regime can be naturally called a *2D helical metal*. In addition to the classical conductivity, σ_c, the carrier diffusion results in a quantum correction, $\delta\sigma$, known as the weak localization correction (Refs. [123–125]). It originates from coherent quantum interference at time scales Δt shorter than

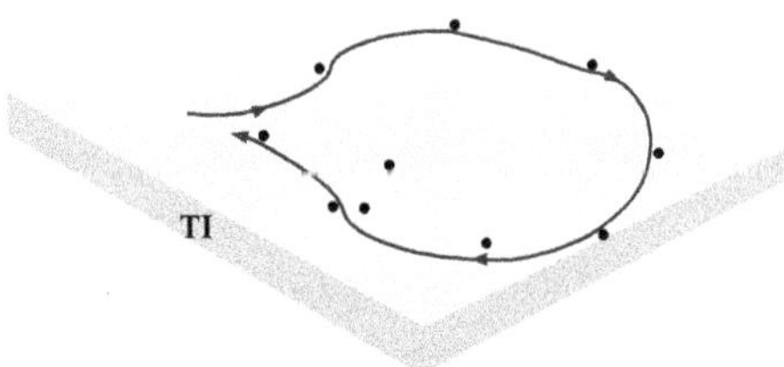

Figure 4.5 Illustration of carrier diffusion via multiple impurity scattering on a TI surface.

a phase-breaking time, τ_ϕ,

$$\Delta t \ll \tau_\phi. \tag{4.45}$$

The studies of such phase-coherent phenomena have emerged as an important tool of characterizing the underlying symmetries of disordered systems, from low-dimensional semiconductors (Refs. [126–134]) to graphene [135–143] and TI materials [144–163]. In view of broad interest in the topic, we intend to present a systematic qualitative and quantitative analysis of the quantum-interference transport for different models of a 2D helical metal.

4.3.1 *Spin–Momentum Locking. Absence of Spin Relaxation in a Helical Metal*

In a helical metal, the direction of the carrier spin $\boldsymbol{\sigma}$ is tide to the direction of its momentum $\boldsymbol{k}$. Such spin–momentum locking is described by the linear terms in the Hamiltonians in Eqs. (4.27), (4.31), and (4.41). These are reminiscent of the Bychkov–Rashba spin–orbit interaction in a conventional 2D electronic system confined in an asymmetric quantum well [164]. The key distinction is the strength of the coupling constant $\mathcal{A}$. Its typical value can be estimated from Eq. (4.17) which gives $\mathcal{A}$ in terms of the Kane matrix element P for the HgTe surface states. For $P = 780$ meV·nm [33, 53] we have $\mathcal{A} \approx 320$ meV·nm. For other materials the experimentally determined coupling constants are $\mathcal{A} = 355$ meV·nm for Bi_2Se_3 [39] and $\mathcal{A} = 255$ meV·nm for Bi_2Te_3 [29, 38]. These numbers are indeed giant compared to the Bychkov–Rashba coupling $\alpha_{BR} \sim 1 \div 10$ meV·nm for conventional GaAs-based electronic systems (Ref. [165]). This large quantitative difference has two important physical consequences. The first is the formation of the conical band dispersion, as shown in Figs. 4.3 and 4.4. The second is the robustness of the spin–momentum locking against impurity scattering (and other spin-independent scattering) as long as the energy separation between the valence and conduction bands is larger than the spectrum broadening $\hbar/\tau$:

$$2\sqrt{\mathcal{A}^2 k_F^2 + \mathcal{M}_{\boldsymbol{k}_F}^2} = 2(E_F - \mathcal{D}k_F^2) \gg \frac{\hbar}{\tau}, \tag{4.46}$$

where τ is the elastic life-time, and k_F is the wave number at which the conduction band crosses the Fermi level. Due to the large constant $\mathcal{A}$ the condition (4.46) can be met already for the Fermi energy values, E_F, of several tens of meV. We note that for the Bychkov–Rashba spin–orbit interaction the condition opposite to Eq. (4.46) is usually assumed,

$$2\alpha_{BR}k_F \ll \frac{\hbar}{\tau}. \tag{4.47}$$

In the latter case, disorder induces interband mixing, enabling *spin relaxation*, that is, the loss of the memory about the initial spin direction after a large number of collisions with impurities. This can occur via D'yakonov–Perel or Elliott–Yafet mechanisms (for the review of the spin relaxation mechanisms, see Ref. [165]).

The condition (4.46) simultaneously implies a metallic transport regime in which

$$2E_F\,\tau/\hbar \approx k_F v\tau_{tr} = k_F\ell_{tr} \gg 1, \tag{4.48}$$

where $\tau_{tr} \approx 2\tau$ and $\ell_{tr} = v\tau_{tr}$ are the transport momentum relaxation time and length, respectively, and v is the Fermi velocity. Thus, the large values of the coupling constant $\mathcal{A}$ enable the metallic regime (4.48) without significant spin relaxation, that is, the carrier spin continues to follow the momentum direction even after a large number of scattering events. Mathematically, this means that even in the presence of impurities the carrier state remains to be an eigenstate of the spin helicity, which for a 2D helical metal is

$$\boldsymbol{\sigma}\cdot\frac{\mathcal{A}\boldsymbol{k}+\mathcal{M}_k\boldsymbol{z}}{|\mathcal{A}\boldsymbol{k}+\mathcal{M}_k\boldsymbol{z}|}, \tag{4.49}$$

where $\boldsymbol{z}$ is the out-of-plane unit vector. This statement is proved rigorously in the following sections.

Spin physics of 2D helical carriers is very attractive for potential applications. In the absence of spin relaxation, information encoded in the carrier spin can be transmitted over large distances, which may open new avenues for *spintronics* in 3DTIs [166]. There has been a growing number of experiments (Refs. [167–174]) and theoretical studies [55, 175–205] covering diverse aspects of the physics of 3DTI materials. Among these are the already mentioned experiments [144–146, 151–153, 155–157, 163] and theoretical papers [147–150, 160–162] addressing the weak localization phenomena. In the next section, we consider a qualitative picture of the weak localization and outline, the differences between the conventional and helical 2D systems.

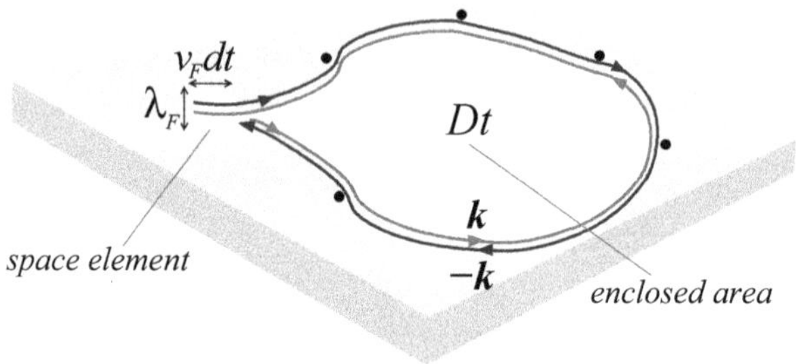

Figure 4.6 Illustration of closed electron trajectories giving rise to weak localization. Regions indicated in white are a space element of the trajectory, $\lambda_F v dt$, and the area enclosed by the trajectory, Dt, which we use to estimate the return probability $d P_{ret}(t)$ in Eq. (4.50).

4.3.2 *Weak (Anti)localization. Qualitative Picture*

At low temperatures a charge carrier can experience many elastic collisions before loosing memory of its original phase. After a series of the scattering events the carrier trajectory can form a closed loop, as depicted in Fig. 4.6. Moreover, if the TRS is preserved, there is an equal probability for the particle to follow the same closed trajectory in the opposite direction. Such closed trajectories cause weak localization (WL) or antilocalization (WAL) depending on whether the spin–orbit coupling is zero or not. In the absence of the spin–orbit coupling, the two counter-propagating states are not protected against backscattering. Their interference yields a negative correction, $\delta\sigma$, to the classical conductivity, σ_c. This is the WL effect, a precursor of the disorder-induced insulating phase. The relative change $\delta\sigma/\sigma_c$ is the measure of the classical return probability integrated from the shortest time-scale, τ, to the largest one, given by the dephasing time τ_φ:

$$\frac{\delta\sigma}{\sigma_c} \sim -\int_\tau^{\tau_\varphi} d P_{ret}(t), \quad d P_{ret}(t) \sim \frac{\lambda_F v dt}{(\sqrt{Dt})^2}. \qquad (4.50)$$

The negative sign here is the consequence of the fact that the interfering states have opposite momenta. The return probability $d P_{ret}(t)$ is proportional to the area of a space element of the trajectory, $\sim \lambda_F v dt$, divided by the typical area, $(\sqrt{Dt})^2$, swept by the particle trajectory after time t (λ_F is the Fermi wavelength and D is the diffusion constant). The conductivity correction (4.50) is logarithmic and has a universal prefactor [123–125]:

$$\delta\sigma = -\frac{2e^2}{2\pi h}\ln\frac{\tau_\varphi}{\tau}, \qquad \tau_\phi \gg \tau, \tag{4.51}$$

where the factor of 2 in the numerator is due to the spin degeneracy. The dephasing time τ_φ depends on the temperature T, increasing for $T \to 0$. The lower the temperature, the more insulating becomes the system.

In contrast to Eq. (4.51), for 2D helical states the experiments and theoretical calculations show that the quantum conductivity correction is, as a rule, positive,

$$\delta\sigma > 0, \tag{4.52}$$

an effect termed *weak antilocalization* (WAL). It indicates that the system remains metallic in the low-temperature limit when $\tau_\varphi(T) \to \infty$. The WAL is intimately related to the absence of backscattering in a helical metal.

Let us take, for example, the TI surface state described by Hamiltonian (4.34). Owing to the spin–momentum locking, the two states on the conjugated trajectories have not only the opposite wave vectors, $\boldsymbol{k}$ and $-\boldsymbol{k}$, but also the opposite spins (indicated by arrows $\rightarrow$ and $\leftarrow$ in Fig. 4.7). These states are related by the TRS operation

$$|-\boldsymbol{k}, \leftarrow\rangle = \mathbb{T}|\boldsymbol{k}, \rightarrow\rangle. \tag{4.53}$$

In this sense, there is a close analogy with the 1D helical edge states considered in Section 2.2. Since the 1D helical states are

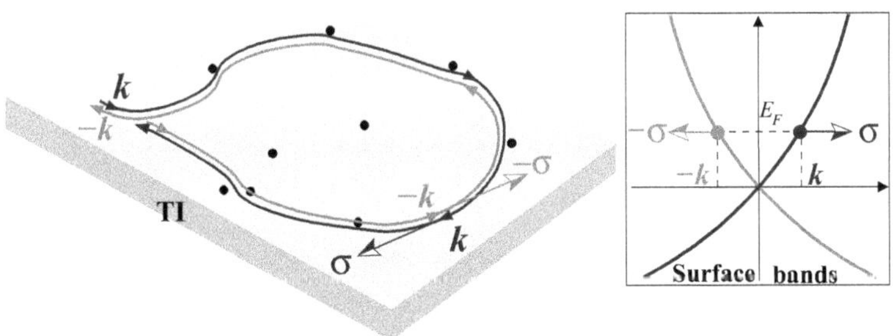

Figure 4.7 Example of closed electron trajectories on a 3DTI surface. Owing to spin–momentum locked band structure (shown in inset), the counter-propagating states are related by TRS and protected against backscattering, leading to WAL instead of WL.

protected against backscattering, the same applies to the TRS-conjugated closed trajectories on a 3DTI surface. The opposite orientation of the spins produces another "minus" sign in Eq. (4.50), yielding the WAL correction

$$\delta\sigma = +\frac{e^2}{2\pi h}\ln\frac{\tau_\varphi}{\tau}. \qquad (4.54)$$

This can also be viewed as the result of the π Berry phase accumulated along the closed loop [135, 136]. The explicit calculation of $\delta\sigma$ with the account of the hexagonal warping under condition (4.46) was carried out in Refs. [148, 160]. We note that the conductivity correction $\delta\sigma$ (4.54) has the same form as for a 2D electron system with spin-orbit impurity scattering [126] or with Bychkov-Rashba and Dresselhaus spin–orbit interactions (Refs. [127, 128]). The reason is that the surface states of 3DTIs with hexagonal warping and conventional 2D electron systems with spin–orbit impurity scattering, Bychkov–Rashba or Dresselhaus spin–orbit interactions belong to the same; *symplectic*, universality class of disordered systems.

A different situation occurs in TI thin films. Let us consider the trajectories of the subsystem described by Hamiltonian H_{k+} in Eq. (4.41), as sketched in Fig. 4.8. The mass term $\mathcal{M}_k\sigma_z$ plays the role of a Zeeman field polarizing the spins out of plane, which is indicated by tilted arrows $\nearrow$ and $\nwarrow$ in Fig. 4.8. In other words, the spins of the counter-propagating states acquire equal out-of-plane components and are no longer antiparallel to each other. These states are, clearly,

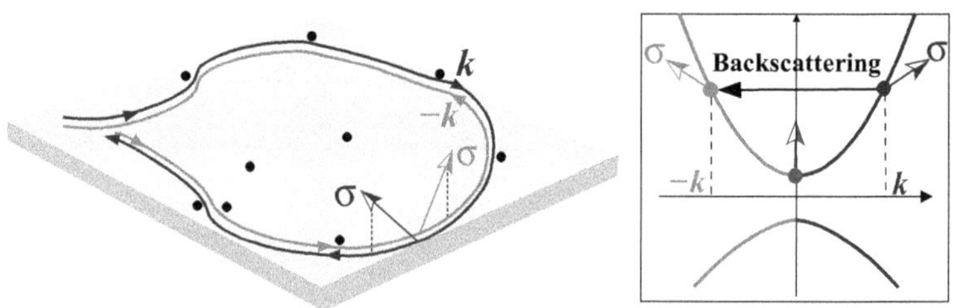

Figure 4.8 Example of closed electron trajectories in a TI thin film. Due to band spin polarization (see inset and Eq. (4.41) for H_+), counter-propagating states with ***k***, $\nearrow$ and $-$***k***, $\nwarrow$ are no longer related by the TRS and experience backscattering, leading to suppressed WAL.

no longer related by the TRS:

$$|-\boldsymbol{k},\nwarrow\rangle \neq \mathbb{T}|\boldsymbol{k},\nearrow\rangle. \tag{4.55}$$

Consequently, the counter-propagating states with $\boldsymbol{k}$ and $-\boldsymbol{k}$ are subject to backscattering. This introduces a new time scale $\tau_{\mathcal{M}}$ for the conductivity correction [147, 148, 150]

$$\delta\sigma_{+} = \frac{e^2}{2\pi h}\ln\frac{\tau^{-1}}{\tau_{\mathcal{M}}^{-1}+\tau_{\varphi}^{-1}}, \qquad \tau_{\mathcal{M}},\ \tau_{\phi} \gg \tau, \tag{4.56}$$

$$\tau_{\mathcal{M}}^{-1} = \frac{2}{\tau}\frac{\mathcal{M}_{k_F}^2}{\mathcal{A}^2 k_F^2 + \mathcal{M}_{k_F}^2}, \qquad k_F = \sqrt{2\pi n}. \tag{4.57}$$

where $\tau_{\mathcal{M}}^{-1}$ is the backscattering rate proportional to the total mass, $\mathcal{M}_{k_F} = \Delta + \mathcal{B}k_F^2$, at the Fermi level, and n is the 2D carrier density in the film.

Since the backscattering rate is temperature independent, the correction (4.56) does not diverge in the $T \to 0$ limit and is, generally, smaller than in the symplectic case, Eq. (4.54). Disordered systems with such suppressed $\delta\sigma$ belong to the *unitary* universality class. That is, the two considered models of a 2D helical metal—a TI surface and a TI thin film—represent two distinct universality classes of disordered systems.

Usually, in the unitary class the TRS is broken. The TI thin films are unique in that they allow one to probe the unitary behavior of the conductivity without the actual TRS breaking. This is because there is the other carrier subsystem [see Eq. (4.41) for H_{-}] whose spin polarization is opposite to that of the $+$ one. The net spin polarization of the $\pm$ subsystems vanishes, whereas their net conductivity is twice that in Eq. (4.56)

$$\delta\sigma = \delta\sigma_{+} + \delta\sigma_{-} = \frac{2e^2}{(2\pi)h}\ln\frac{\tau^{-1}}{\tau_{\mathcal{M}}^{-1}+\tau_{\varphi}^{-1}}. \tag{4.58}$$

Figure 4.9 shows the typical dependence of $\delta\sigma$ on carrier density n and band gap Δ. The next section contains the details of the calculation of the quantum conductivity correction.

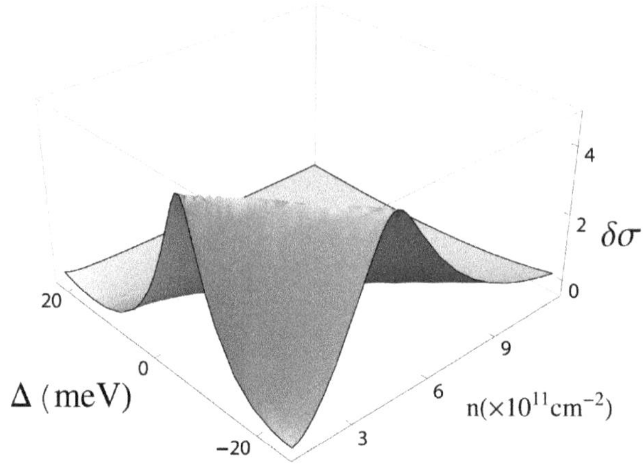

Figure 4.9 WAL conductivity correction $\delta\sigma$ [in units of $e^2/\pi h$, see Eq. (4.58)] versus carrier density n and band gap Δ (adapted from Ref. [148]).

4.4 Weak Antilocalization in TI thin Films

4.4.1 *Kubo Formula. Preliminaries*

We begin with the TI thin-film model whose $\boldsymbol{k}$-space isotropy makes the calculations more transparent. The general formula for the electric conductivity in the linear approximation with respect to an external uniform electric field $\mathcal{E}e^{-i\omega t}$ is given by Kubo's theory (Ref. [47]). We are interested in the longitudinal conductivity σ_{xx} that can be written as

$$\sigma_{xx} = \frac{e^2}{2\pi\omega a}\int dE[f(E) - f(E+\hbar\omega)] \times \sum_{\boldsymbol{k},\boldsymbol{k}'} \mathrm{Tr}[\hat{v}^x_{\boldsymbol{k}} \hat{G}^R_{\boldsymbol{k},\boldsymbol{k}'}(E+\hbar\omega)\hat{v}^x_{\boldsymbol{k}'}\hat{G}^A_{\boldsymbol{k}',\boldsymbol{k}}(E)], \tag{4.59}$$

where a is the area of the system, $f(E)$ is the Fermi–Dirac distribution function, Tr denotes the trace in spin space, $\hat{v}^x_{\boldsymbol{k}}$ is the x-component of the velocity operator

$$\hat{\boldsymbol{v}}_{\boldsymbol{k}} = \hbar^{-1}\partial_{\boldsymbol{k}}\hat{H}_{\boldsymbol{k}}, \tag{4.60}$$

and $\hat{G}^{R/A}_{\boldsymbol{k},\boldsymbol{k}'}(E)$ is the retarded/advanced Green function satisfying the equation

$$\sum_{\boldsymbol{k},\boldsymbol{k}_1}[(E_{R/A} - \hat{H}_{\boldsymbol{k}})\delta_{\boldsymbol{k},\boldsymbol{k}_1} - \hat{V}_{\boldsymbol{k},\boldsymbol{k}_1}]\hat{G}^{R/A}_{\boldsymbol{k}_1,\boldsymbol{k}'} = \delta_{\boldsymbol{k},\boldsymbol{k}'}\sigma_0, \tag{4.61}$$

where energy $E_{R/A} = E \pm i0$ includes an infinitesimal imaginary part, $i0$, with $\pm$ for the retarded and advanced functions, respectively. In Eqs. (4.60) and (4.61), $\hat{H}_{\boldsymbol{k}}$ is the Hamiltonian of the system in the absence of any inhomogeneities, while $\hat{V}_{\boldsymbol{k},\boldsymbol{k}'}$ is the matrix element of an inhomogeneous potential $\hat{V}(\boldsymbol{r})$, defined by

$$\hat{V}_{\boldsymbol{k},\boldsymbol{k}'} = a^{-1} \int \mathrm{d}r\, \hat{V}(\boldsymbol{r}) e^{-i(\boldsymbol{k}-\boldsymbol{k}')\boldsymbol{r}}. \tag{4.62}$$

In the above equations and throughout the "hat" indicates 2×2 matrices in spin space.

For arbitrary potential $\hat{V}(\boldsymbol{r})$, the solution of Eq. (4.61) can be represented by an infinite series expansion in powers of matrix elements $\hat{V}_{\boldsymbol{k},\boldsymbol{k}'}$. For that purpose, we recast Eq. (4.61) in the integral form

$$\hat{G}^{R/A}_{\boldsymbol{k},\boldsymbol{k}'} = \hat{G}^{R/A}_{0\boldsymbol{k}} \delta_{\boldsymbol{k},\boldsymbol{k}'} + \sum_{\boldsymbol{k}_1} \hat{G}^{R/A}_{0\boldsymbol{k}} \hat{V}_{\boldsymbol{k},\boldsymbol{k}_1} \hat{G}^{R/A}_{\boldsymbol{k}_1,\boldsymbol{k}'}, \tag{4.63}$$

where $\hat{G}^{R/A}_{0\boldsymbol{k}}$ is the Green function of the homogeneous system, obeying the equation

$$(E_{R/A} - \hat{H}_{\boldsymbol{k}}) \hat{G}^{R/A}_{0\boldsymbol{k}} = \sigma_0. \tag{4.64}$$

Iterating Eq. (4.63), we generate the series

$$\hat{G}^{R/A}_{\boldsymbol{k},\boldsymbol{k}'} = \hat{G}^{R/A}_{0\boldsymbol{k}} \delta_{\boldsymbol{k},\boldsymbol{k}'} + \hat{G}^{R/A}_{0\boldsymbol{k}} \hat{V}_{\boldsymbol{k},\boldsymbol{k}'} \hat{G}^{R/A}_{0\boldsymbol{k}'} \tag{4.65}$$

$$+ \sum_{\boldsymbol{k}_1} \hat{G}^{R/A}_{0\boldsymbol{k}} \hat{V}_{\boldsymbol{k},\boldsymbol{k}_1} \hat{G}^{R/A}_{0\boldsymbol{k}_1} \hat{V}_{\boldsymbol{k}_1,\boldsymbol{k}'} \hat{G}^{R/A}_{0\boldsymbol{k}'} \tag{4.66}$$

$$+ \sum_{\boldsymbol{k}_1,\boldsymbol{k}_2} \hat{G}^{R/A}_{0\boldsymbol{k}} \hat{V}_{\boldsymbol{k},\boldsymbol{k}_1} \hat{G}^{R/A}_{0\boldsymbol{k}_1} \hat{V}_{\boldsymbol{k}_1,\boldsymbol{k}_2} \hat{G}^{R/A}_{0\boldsymbol{k}_2} \hat{V}_{\boldsymbol{k}_2,\boldsymbol{k}'} \hat{G}^{R/A}_{0\boldsymbol{k}'} \tag{4.67}$$

$$+ \sum_{\boldsymbol{k}_1,\boldsymbol{k}_2,\boldsymbol{k}_3} \hat{G}^{R/A}_{0\boldsymbol{k}} \hat{V}_{\boldsymbol{k},\boldsymbol{k}_1} \hat{G}^{R/A}_{0\boldsymbol{k}_1} \hat{V}_{\boldsymbol{k}_1,\boldsymbol{k}_2} \hat{G}^{R/A}_{0\boldsymbol{k}_2} \hat{V}_{\boldsymbol{k}_2,\boldsymbol{k}_3} \hat{G}^{R/A}_{0\boldsymbol{k}_3} \hat{V}_{\boldsymbol{k}_3,\boldsymbol{k}'} \hat{G}^{R/A}_{0\boldsymbol{k}'} + \ldots \tag{4.68}$$

It is worth emphasizing that this is a matrix series in which different entries do not generally commute. Figure 4.10 gives a useful diagrammatic representation of the solution in Eq. (4.65). It allows one to easily generate expansion terms of any required order.

In the equations above, the specifics of the system enters through Hamiltonian $\hat{H}_{\boldsymbol{k}}$ and Green function $\hat{G}^{R/A}_{0\boldsymbol{k}}$. We now apply these general

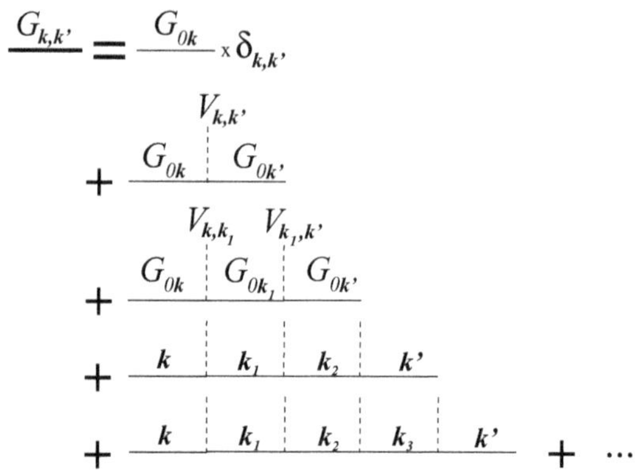

Figure 4.10 Diagrammatic representation of Green function (4.65): the thick solid line corresponds to the exact Green function $\hat{G}^{R/A}_{k,k'}$, a thin solid line depicts the unperturbed Green function $\hat{G}^{R/A}_{0k}$, and a vertical dashed line represents the potential matrix element $\hat{V}_{k,k'}$. Summation is assumed over all the intermediate wave vectors $\boldsymbol{k}_1, \boldsymbol{k}_2, \boldsymbol{k}_3, \ldots$

formulae to a TI thin film. Since its Hamiltonian (4.41) is block-diagonal, it suffices to consider only the + block, that is, Hamiltonian $\hat{H}_{\boldsymbol{k}}$ in Eqs. (4.59)–(4.62) is identified as $\hat{H}_{\boldsymbol{k}+}$ from Eq. (4.41),

$$\hat{H}_{\boldsymbol{k}} = \hat{H}_{\boldsymbol{k}+}. \tag{4.69}$$

In this case, the velocity operator (4.60) is

$$\hat{\boldsymbol{v}}_{\boldsymbol{k}} = \hbar^{-1}(\mathcal{A}\boldsymbol{\sigma} + \partial_{\boldsymbol{k}}\mathcal{M}_{\boldsymbol{k}}\sigma_z + 2\mathcal{D}\boldsymbol{k}\sigma_0) \tag{4.70}$$

$$= \hbar^{-1}(\mathcal{A}\boldsymbol{\sigma} + 2\mathcal{B}\boldsymbol{k}\sigma_z + 2\mathcal{D}\boldsymbol{k}\sigma_0), \tag{4.71}$$

and Eq. (4.64) for the unperturbed Green function $\hat{G}^{R/A}_{0\boldsymbol{k}}$ reads

$$(E_{R/A} + E_F - \mathcal{D}\boldsymbol{k}^2 - \mathcal{A}\boldsymbol{\sigma}\cdot\boldsymbol{k} - \mathcal{M}_{\boldsymbol{k}}\sigma_z)\hat{G}^{R/A}_{0\boldsymbol{k}} = \sigma_0. \tag{4.72}$$

In order to invert this equation, we apply matrix $E_{R/A} + E_F - \mathcal{D}\boldsymbol{k}^2 + \mathcal{A}\boldsymbol{\sigma}\cdot\boldsymbol{k} + \mathcal{M}_{\boldsymbol{k}}\sigma_z$ from the left:

$$\hat{G}^{R/A}_{0\boldsymbol{k}} = \frac{E_{R/A} + E_F - \mathcal{D}\boldsymbol{k}^2 + \mathcal{A}\boldsymbol{\sigma}\cdot\boldsymbol{k} + \mathcal{M}_{\boldsymbol{k}}\sigma_z}{[E_{R/A} - \xi_{\boldsymbol{k}}][E_{R/A} - \bar{\xi}_{\boldsymbol{k}}]}, \tag{4.73}$$

where

$$\xi_{\boldsymbol{k}} = \sqrt{\mathcal{A}^2\boldsymbol{k}^2 + \mathcal{M}^2_{\boldsymbol{k}}} + \mathcal{D}\boldsymbol{k}^2 - E_F, \quad \bar{\xi}_{\boldsymbol{k}} = -\sqrt{\mathcal{A}^2\boldsymbol{k}^2 + \mathcal{M}^2_{\boldsymbol{k}}} + \mathcal{D}\boldsymbol{k}^2 - E_F, \tag{4.74}$$

are the conduction- and valence-band dispersions, respectively.

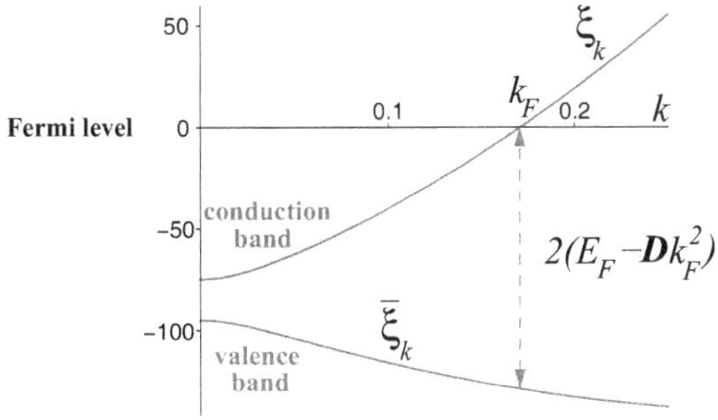

Figure 4.11 Conduction- and valence-band energies (in meV) versus wave number k (in nm^{-1}) for typical parameters $\mathcal{A} = 350$ meV nm, $\mathcal{B} = 500$ meV nm^2, $\mathcal{D} = 700$ meV nm^2, $\mathcal{M} = 10$ meV, and $E_F = 85$ meV; see Eq. (4.74).

We assume that the Fermi level lies in the conduction band, as depicted in Figs. 4.4 and 4.11. The corresponding Fermi surface is a closed line in momentum space, which is determined by

$$\xi_{\boldsymbol{k}_F} = \sqrt{\mathcal{A}^2 \boldsymbol{k}_F^2 + \mathcal{M}_{\boldsymbol{k}_F}^2} + \mathcal{D}\boldsymbol{k}_F^2 - E_F = 0. \tag{4.75}$$

This is an equation for the Fermi wave vector, $\boldsymbol{k}_F$. At the same time, the valence band energy $\overline{\xi}_{\boldsymbol{k}}$ does not vanish for $\boldsymbol{k} = \boldsymbol{k}_F$, being always negative (see also Fig. 4.11):

$$\overline{\xi}_{\boldsymbol{k}_F} = -\sqrt{\mathcal{A}^2 \boldsymbol{k}_F^2 + \mathcal{M}_{\boldsymbol{k}_F}^2} + \mathcal{D}\boldsymbol{k}_F^2 - E_F \tag{4.76}$$

$$= -2\sqrt{\mathcal{A}^2 \boldsymbol{k}_F^2 + \mathcal{M}_{\boldsymbol{k}_F}^2} \tag{4.77}$$

$$= -2(E_F - \mathcal{D}\boldsymbol{k}_F^2). \tag{4.78}$$

The difference $\xi_{\boldsymbol{k}_F} - \overline{\xi}_{\boldsymbol{k}_F}$ defines the energy separation between the conduction and valence bands for given $\boldsymbol{k}_F$,

$$\xi_{\boldsymbol{k}_F} - \overline{\xi}_{\boldsymbol{k}_F} = 2\sqrt{\mathcal{A}^2 \boldsymbol{k}_F^2 + \mathcal{M}_{\boldsymbol{k}_F}^2} = 2(E_F - \mathcal{D}\boldsymbol{k}_F^2). \tag{4.79}$$

Since in many situations only the wave vectors close to $\boldsymbol{k}_F$ matter, we can simplify Eq. (4.73) as follows

$$\hat{G}_{0\boldsymbol{k}}^{R/A} \approx \frac{E_{R/A} + \sqrt{\mathcal{A}^2 \boldsymbol{k}_F^2 + \mathcal{M}_{\boldsymbol{k}_F}^2} + \mathcal{A}\boldsymbol{\sigma} \cdot \boldsymbol{k}_F + \mathcal{M}_{\boldsymbol{k}_F}\sigma_z}{[E_{R/A} - \xi_{\boldsymbol{k}}][E_{R/A} + 2\sqrt{\mathcal{A}^2 \boldsymbol{k}_F^2 + \mathcal{M}_{\boldsymbol{k}_F}^2}]}, \quad \boldsymbol{k} \approx \boldsymbol{k}_F. \tag{4.80}$$

Also, for sufficiently large E_F the band separation (4.79) can be made larger than the characteristic energy scale of $E_{R/A}$, i.e.

$$2\sqrt{\mathcal{A}^2\boldsymbol{k}_F^2 + \mathcal{M}_{\boldsymbol{k}_F}^2} = 2(E_F - \mathcal{D}\boldsymbol{k}_F^2) \gg E_{R/A}. \tag{4.81}$$

In a disordered system, $E_{R/A} \sim \hbar/\tau$, where τ is the elastic lifetime [*cf.* Eq. (4.46)]. Under condition (4.81), the Green function (4.80) assumes a rather compact form

$$\hat{G}_{0\boldsymbol{k}}^{R/A} \approx \frac{\hat{P}_{\boldsymbol{n}}}{E_{R/A} - \xi_{\boldsymbol{k}}}, \qquad \xi_{\boldsymbol{k}} \approx \hbar v(|\boldsymbol{k}| - k_F), \tag{4.82}$$

$$\hat{P}_{\boldsymbol{n}} = \frac{\sigma_0 + \boldsymbol{\sigma}\cdot\mathbf{e}_{\boldsymbol{n}}}{2}, \qquad \mathbf{e}_{\boldsymbol{n}} = \frac{\mathcal{A}k_F\boldsymbol{n} + \mathcal{M}_{k_F}\boldsymbol{z}}{\sqrt{\mathcal{A}^2k_F^2 + \mathcal{M}_{k_F}^2}}, \qquad \boldsymbol{n} = \frac{\boldsymbol{k}}{|\boldsymbol{k}|}, \tag{4.83}$$

where the information about the carrier spin is contained in the matrix $\hat{P}_{\boldsymbol{n}}$. In particular, the unit vector $\mathbf{e}_{\boldsymbol{n}}$ indicates the direction of the carrier spin for given direction of the momentum, $\boldsymbol{n}$ ($\boldsymbol{z} \perp \boldsymbol{n}$ is the unit out-of-plane vector). That is, $\boldsymbol{\sigma}\cdot\mathbf{e}_{\boldsymbol{n}}$ is the spin helicity matrix, which is of primary importance throughout. It follows from Eq. (4.82) that

$$(\sigma\cdot\mathbf{e}_{\boldsymbol{n}})\hat{G}_{0\boldsymbol{k}}^{R/A} = \hat{G}_{0\boldsymbol{k}}^{R/A}, \tag{4.84}$$

meaning that Green function (4.82) describes a state with the helicity eigenvalue equal to $+1$. It can be shown that the other eigenvalue, -1, corresponds to the valence band states. Since the states near the Fermi surface $|\boldsymbol{k}| \approx k_F$ matter most, the conduction band dispersion, $\xi_{\boldsymbol{k}}$, in Eq. (4.82) is linearized near k_F.

4.4.2 *Potential Disorder*

We narrow further our model by considering a spin-independent potential of the form

$$\hat{V}(\boldsymbol{r}) = V(\boldsymbol{r})\sigma_0, \tag{4.85}$$

where σ_0 denotes a 2×2 unit spin matrix, and $V(\boldsymbol{r})$ is a random function of position $\boldsymbol{r}$. We assume that the disorder originates from a large number of randomly distributed impurities with a homogeneous density in the thermodynamic limit. In this case, $V(\boldsymbol{r})$

has a Gaussian distribution which is fully characterized by a two-point correlation function. We assume short-ranged correlations modeled by

$$\zeta(\boldsymbol{r}-\boldsymbol{r}') = \langle\langle V(\mathbf{r})V(\boldsymbol{r}')\rangle\rangle = \frac{\hbar}{\pi N\tau_0}\,\delta(\boldsymbol{r}-\boldsymbol{r}'), \tag{4.86}$$

where $\langle\langle\ldots\rangle\rangle$ denotes averaging over the random positions of the impurities. For convenience of further calculations, the correlation function is parametrized in terms of the characteristic scattering time, τ_0, and the density of states (DOS) at the Fermi level,

$$N = \frac{k_F}{2\pi\hbar v}. \tag{4.87}$$

From Eqs. (4.62) and (4.85) we derive the correlation function for the matrix elements of the random potential,

$$\langle\langle \hat{V}_{\boldsymbol{k},\boldsymbol{k}_1} \otimes \hat{V}_{\boldsymbol{k}',\boldsymbol{k}_2}\rangle\rangle = \sigma_0 \otimes \sigma_0 \tag{4.88}$$

$$\times \int \frac{d\boldsymbol{r}_1 d\boldsymbol{r}_2}{a^2}\zeta(\boldsymbol{r}_1-\boldsymbol{r}_2)e^{-i\mathbf{r}_1(\boldsymbol{k}-\boldsymbol{k}_1)-i\mathbf{r}_2(\boldsymbol{k}'-\boldsymbol{k}_2)} \tag{4.89}$$

$$= \sigma_0 \otimes \sigma_0 \int \frac{d\boldsymbol{r}d\boldsymbol{R}}{a^2}\zeta(\mathbf{r})e^{-i\mathbf{r}(\boldsymbol{k}-\boldsymbol{k}_1)-i\boldsymbol{R}(\boldsymbol{k}-\boldsymbol{k}_1+\boldsymbol{k}'-\boldsymbol{k}_2)} \tag{4.90}$$

$$= \frac{\zeta_{\boldsymbol{k}-\boldsymbol{k}_1}}{a}\delta_{\boldsymbol{k}-\boldsymbol{k}_1,-\boldsymbol{k}'+\boldsymbol{k}_2}\sigma_0 \otimes \sigma_0, \tag{4.91}$$

where $\zeta_{\boldsymbol{k}}$ is the Fourier transform of the correlation function $\zeta(\boldsymbol{r})$,

$$\zeta_{\boldsymbol{k}} = \int \zeta(\mathbf{r})\mathrm{e}^{-i\boldsymbol{k}\mathbf{r}}d\mathbf{r} \equiv \zeta = \frac{\hbar}{\pi N\tau_0}, \tag{4.92}$$

and $\sigma_0 \otimes \sigma_0$ indicates the direct product of the unit matrices. We will also use an alternative notation

$$\langle\langle V_{\alpha\alpha_1}(\boldsymbol{k},\boldsymbol{k}_1)V_{\beta\beta_2}(\boldsymbol{k}',\boldsymbol{k}_2)\rangle\rangle = \frac{\zeta}{a}\delta_{\boldsymbol{k}-\boldsymbol{k}_1,-\boldsymbol{k}'+\boldsymbol{k}_2}\delta_{\alpha,\alpha_1}\delta_{\beta,\beta_2}, \tag{4.93}$$

where α and β are the spin indices. The wave-vector selection rule is quite simple: the difference between the first two wave vectors plus that between the second two wave vectors equals to zero. The higher-even-order correlators are reduced to products of the second-order ones through pair contractions, for example,

$$\langle\langle \hat{V}_{\boldsymbol{k},\boldsymbol{k}_1} \otimes \hat{V}_{\boldsymbol{k}',\boldsymbol{k}_2} \otimes \hat{V}_{\boldsymbol{k}'',\boldsymbol{k}_3} \otimes \hat{V}_{\boldsymbol{k}''',\boldsymbol{k}_4}\rangle\rangle \tag{4.94}$$

$$= \hat{V}_{\boldsymbol{k},\boldsymbol{k}_1} \otimes \hat{V}_{\boldsymbol{k}',\boldsymbol{k}_2} \otimes \hat{V}_{\boldsymbol{k}'',\boldsymbol{k}_3} \otimes \hat{V}_{\boldsymbol{k}''',\boldsymbol{k}_4}$$

$$+ \hat{V}_{\boldsymbol{k},\boldsymbol{k}_1} \otimes \hat{V}_{\boldsymbol{k}',\boldsymbol{k}_2} \otimes \hat{V}_{\boldsymbol{k}'',\boldsymbol{k}_3} \otimes \hat{V}_{\boldsymbol{k}''',\boldsymbol{k}_4}$$

$$+ \hat{V}_{\boldsymbol{k},\boldsymbol{k}_1} \otimes \hat{V}_{\boldsymbol{k}',\boldsymbol{k}_2} \otimes \hat{V}_{\boldsymbol{k}'',\boldsymbol{k}_3} \otimes \hat{V}_{\boldsymbol{k}''',\boldsymbol{k}_4},$$

where we indicate all possible pair correlators. The odd correlation functions for a Gaussian random process vanish

$$\langle\langle \hat{V}_{\mathbf{k},\mathbf{k}_1} \rangle\rangle = 0, \qquad \langle\langle \hat{V}_{\mathbf{k},\mathbf{k}_1} \otimes \hat{V}_{\mathbf{k}',\mathbf{k}_2} \otimes \hat{V}_{\mathbf{k}'',\mathbf{k}_3} \rangle\rangle = 0, \qquad \ldots \tag{4.95}$$

4.4.3 *Nonperturbative Treatment of Disorder. Diagrammatics*

The strategy to calculate the conductivity is conceptually simple. We should insert the series expansion (4.65)–(4.68) for the Green functions into Kubo formula (4.59), and average, the latter over disorder configurations. On this way we encounter two types of averaged Green functions: the retarded and advanced function averaged independently, that is

$$\langle\langle \hat{G}^R_{\mathbf{k},\mathbf{k}'}(E + \hbar\omega) \rangle\rangle, \qquad \langle\langle \hat{G}^A_{\mathbf{k},\mathbf{k}'} \rangle\rangle, \tag{4.96}$$

and the average direct product of the retarded and advanced functions

$$\langle\langle \hat{G}^R_{\mathbf{k},\mathbf{k}'}(E + \hbar\omega) \otimes \hat{G}^A_{\mathbf{k}',\mathbf{k}} \rangle\rangle. \tag{4.97}$$

We begin by calculating the Green functions (4.96). The first step is to average each term in the expansion (4.65) with the help of the Gaussian correlators from Eqs. (4.91)–(4.95). Diagrammatically, this amounts to connecting pairs of the dashes lines in Fig. 4.10 in all possible ways, as depicted in Fig. 4.12 (the superscripts R/A have been omitted for clarity). Only the terms with even powers of the disorder potential remain

$$\langle\langle \hat{G}_{\mathbf{k},\mathbf{k}'} \rangle\rangle = \hat{G}^{(0)}_{\mathbf{k},\mathbf{k}'} + \hat{G}^{(2)}_{\mathbf{k},\mathbf{k}'} + \hat{G}^{(4)}_{\mathbf{k},\mathbf{k}'} + \ldots \tag{4.98}$$

where $\hat{G}^{(0)}_{\mathbf{k},\mathbf{k}'}$, $\hat{G}^{(2)}_{\mathbf{k},\mathbf{k}'}$, and $\hat{G}^{(4)}_{\mathbf{k},\mathbf{k}'}$ are represented by the zeroth-, second-, and fourth-order diagrams in Fig. 4.12. The corresponding analytical expressions are

$$\hat{G}^{(0)}_{\mathbf{k},\mathbf{k}'} = \hat{G}_{0\mathbf{k}}\delta_{\mathbf{k},\mathbf{k}'}, \tag{4.99}$$

$$\begin{aligned} \hat{G}^{(2)}_{\mathbf{k},\mathbf{k}'} &= \hat{G}_{0\mathbf{k}} \sum_{\mathbf{k}_1} \underbrace{\hat{V}_{\mathbf{k},\mathbf{k}_1} \hat{G}_{0\mathbf{k}_1} \hat{V}_{\mathbf{k}_1,\mathbf{k}'}}\hat{G}_{0\mathbf{k}'} = \hat{G}_{0\mathbf{k}}\delta_{\mathbf{k},\mathbf{k}'}\frac{\zeta}{a}\sum_{\mathbf{k}_1} \hat{G}_{0\mathbf{k}_1}\hat{G}_{0\mathbf{k}} \\ &= \hat{G}^{(0)}_{\mathbf{k},\mathbf{k}'}\zeta \int \frac{d\mathbf{k}_1}{(2\pi)^2}\hat{G}_{0\mathbf{k}_1}\hat{G}_{0\mathbf{k}}, \end{aligned} \tag{4.100}$$

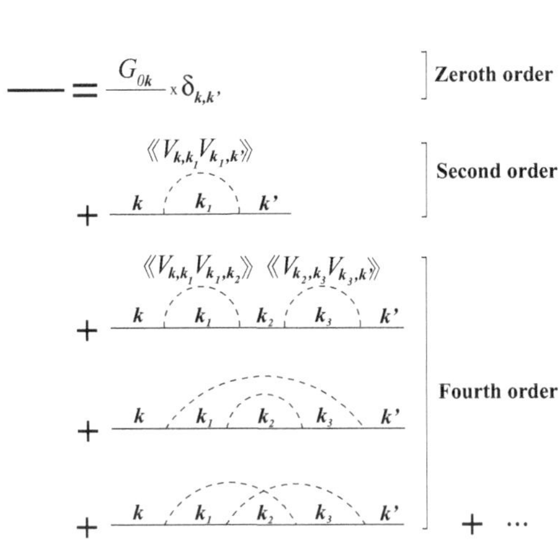

Figure 4.12 Diagrammatic representation of averaged Green function (4.65) up to fourth order in disorder potential.

$$\hat{G}^{(4)}_{\boldsymbol{k},\boldsymbol{k}'} = \hat{G}_{0\boldsymbol{k}} \sum_{\boldsymbol{k}_1,\boldsymbol{k}_2,\boldsymbol{k}_3} \hat{V}_{\boldsymbol{k},\boldsymbol{k}_1} \hat{G}_{0\boldsymbol{k}_1} \hat{V}_{\boldsymbol{k}_1,\boldsymbol{k}_2} \hat{G}_{0\boldsymbol{k}_2} \hat{V}_{\boldsymbol{k}_2,\boldsymbol{k}_3} \hat{G}_{0\boldsymbol{k}_3} \hat{V}_{\boldsymbol{k}_3,\boldsymbol{k}'} \hat{G}_{0\boldsymbol{k}'} \tag{4.101}$$

$$+\hat{G}_{0\boldsymbol{k}} \sum_{\boldsymbol{k}_1,\boldsymbol{k}_2,\boldsymbol{k}_3} \hat{V}_{\boldsymbol{k},\boldsymbol{k}_1} \hat{G}_{0\boldsymbol{k}_1} \hat{V}_{\boldsymbol{k}_1,\boldsymbol{k}_2} \hat{G}_{0\boldsymbol{k}_2} \hat{V}_{\boldsymbol{k}_2,\boldsymbol{k}_3} \hat{G}_{0\boldsymbol{k}_3} \hat{V}_{\boldsymbol{k}_3,\boldsymbol{k}'} \hat{G}_{0\boldsymbol{k}'} \tag{4.102}$$

$$+\hat{G}_{0\boldsymbol{k}} \sum_{\boldsymbol{k}_1,\boldsymbol{k}_2,\boldsymbol{k}_3} \hat{V}_{\boldsymbol{k},\boldsymbol{k}_1} \hat{G}_{0\boldsymbol{k}_1} \hat{V}_{\boldsymbol{k}_1,\boldsymbol{k}_2} \hat{G}_{0\boldsymbol{k}_2} \hat{V}_{\boldsymbol{k}_2,\boldsymbol{k}_3} \hat{G}_{0\boldsymbol{k}_3} \hat{V}_{\boldsymbol{k}_3,\boldsymbol{k}'} \hat{G}_{0\boldsymbol{k}'}. \tag{4.103}$$

Evaluating the indicated pair correlators with the help of Eq. (4.93), we have

$$\hat{G}^{(4)}_{\boldsymbol{k},\boldsymbol{k}'} = \hat{G}^{(0)}_{\boldsymbol{k},\boldsymbol{k}'} \left[\zeta \int \frac{d\boldsymbol{k}_1}{(2\pi)^2} \hat{G}_{0\boldsymbol{k}_1}\right] \hat{G}_{0\boldsymbol{k}} \left[\zeta \int \frac{d\boldsymbol{k}_3}{(2\pi)^2} \hat{G}_{0\boldsymbol{k}_3}\right] \hat{G}_{0\boldsymbol{k}} \tag{4.104}$$

$$+ \hat{G}^{(0)}_{\boldsymbol{k},\boldsymbol{k}'} \left[\zeta \int \frac{d\boldsymbol{k}_1}{(2\pi)^2} \hat{G}_{0\boldsymbol{k}_1} \left[\zeta \int \frac{d\boldsymbol{k}_2}{(2\pi)^2} \hat{G}_{0\boldsymbol{k}_2}\right] \hat{G}_{0\boldsymbol{k}_1}\right] \hat{G}_{0\boldsymbol{k}} \tag{4.105}$$

$$+ \hat{G}^{(0)}_{\boldsymbol{k},\boldsymbol{k}'} \left[\zeta \int \frac{d\boldsymbol{k}_1}{(2\pi)^2} \hat{G}_{0\boldsymbol{k}_1} \left[\zeta \int \frac{d\boldsymbol{k}_2}{(2\pi)^2} \hat{G}_{0\boldsymbol{k}_2} \hat{G}_{0\boldsymbol{k}_2-\boldsymbol{k}_1+\boldsymbol{k}}\right]\right] \hat{G}_{0\boldsymbol{k}}. \tag{4.106}$$

The obtained series (4.98) is not a perturbative expansion of powers of a certain small parameter. Generally, there is no small parameter because the Green function $\hat{G}_{0\boldsymbol{k}}$ (4.82) diverges at the Fermi surface $|\boldsymbol{k}| = k_F$ for $E \to 0$. It is quite obvious from

Eq. (4.100) for the second-order term $\hat{G}^{(2)}_{\boldsymbol{k},\boldsymbol{k}'}$. As we cannot truncate the series at any finite order, the usual perturbative treatment fails.

At the same time, it is not possible to calculate the infinite sum exactly. The way out of this problem is to eliminate *irrelevant* terms and then try to sum up the rest of the infinite series by analogy with the Feynman diagrammatic technique. The irrelevant terms are those that *do* contain a small parameter compared with the previous order. As we just saw, this does not apply to the second-order term $\hat{G}^{(2)}_{\boldsymbol{k},\boldsymbol{k}'}$. However, an irrelevant term appears in the next, fourth-order. This is the last (crossed) diagram in Fig. 4.12. In its analytical expression (4.106), the $\boldsymbol{k}_2$ integration involves two Green functions,

$$\hat{G}_{0\boldsymbol{k}_2}$$

and

$$\hat{G}_{0\boldsymbol{k}_2-\boldsymbol{k}_1+\boldsymbol{k}}.$$

Due to the wave-vector difference, these functions cannot be simultaneously singular on the Fermi surface, leading to a suppressed contribution. To estimate it, let us consider the integral

$$\zeta \int \frac{d\boldsymbol{k}_2}{(2\pi)^2} \hat{G}_{0\boldsymbol{k}_2} \hat{G}_{0\boldsymbol{k}_2-\boldsymbol{k}_1+\boldsymbol{k}} \approx \tag{4.107}$$

$$\frac{\zeta k_F}{2\pi} \int\limits_{-\infty}^{\infty} \frac{d\kappa}{E - \hbar v \kappa} \overline{\frac{\hat{P}_{\boldsymbol{n}_2} \hat{P}_{\boldsymbol{n}_2-\boldsymbol{n}_1+\boldsymbol{n}}}{E - \hbar v(|(\kappa + k_F)\boldsymbol{n}_2 - \boldsymbol{k}_1 + \boldsymbol{k}| - k_F)}}, \tag{4.108}$$

where we switched to the integration over the wave-vector difference $\kappa = k_2 - k_F$, extending the lower integration limit from $-k_F$ to $-\infty$, which is permitted under condition (4.81), and the bar denotes averaging over the directions of the unit vector $\boldsymbol{n}_2$,

$$\overline{(\ldots)} = \int_0^{2\pi} \frac{d\phi_{\boldsymbol{n}_2}}{2\pi} \ldots \tag{4.109}$$

Since the function in Eq. (4.108) is integrable for $\kappa \to \pm\infty$, the integration path can chosen around the pole of the first fraction, which yields

$$\pm 2\pi i \zeta N \overline{\frac{\hat{P}_{\boldsymbol{n}_2} \hat{P}_{\boldsymbol{n}_2-\boldsymbol{n}_1+\boldsymbol{n}}}{\hbar v(|(\frac{E}{\hbar v} + k_F)\boldsymbol{n}_2 - \boldsymbol{k}_1 + \boldsymbol{k}| - k_F) - E}}, \tag{4.110}$$

where $\pm$ are for the retarded and advanced functions, respectively. The wave vectors $\boldsymbol{k}$ and $\boldsymbol{k}_1$ can be taken on the Fermi surface,

$$\boldsymbol{k} = k_F \boldsymbol{n}, \qquad \boldsymbol{k}_1 = k_F \boldsymbol{n}_1, \tag{4.111}$$

because they are the arguments of two different Green functions, $\hat{G}_{0\boldsymbol{k}}$ and $\hat{G}_{0\boldsymbol{k}_1}$, each being singular on the Fermi surface [see Eq. (4.106)]. For $\hbar v k_F \gg E$ and ζ given by Eq. (4.92), the estimate of Eq. (4.110) is

$$\pm \frac{2i}{k_F v \tau_0} \overline{\frac{\hat{P}_{\boldsymbol{n}_2} \hat{P}_{\boldsymbol{n}_2 - \boldsymbol{n}_1 + \boldsymbol{n}}}{|\boldsymbol{n}_2 - \boldsymbol{n}_1 + \boldsymbol{n}| - 1}}. \tag{4.112}$$

The prefactor is a small parameter $\sim 1/k_F v \tau_0 \sim 1/k_F \ell_{tr} \ll 1$ [*cf.*, Eq. (4.48)]. Although the denominator $|\boldsymbol{n}_2 - \boldsymbol{n}_1 + \boldsymbol{n}| - 1$ can vanish, this cannot compensate the smallness of the prefactor because this happens only for the special direction of $\boldsymbol{n}_2 = \boldsymbol{n}_1 - \boldsymbol{n}$, which has a small weight in the angle integral. Therefore, in the metallic regime with $k_F \ell_{tr} \gg 1$ the contribution of the crossed diagram [see Eq. (4.106)] is small compared to the second-order term (4.100),

$$\hat{G}^{(4)}_{\boldsymbol{k},\boldsymbol{k}'}|_{\text{crossed}} \sim \frac{1}{k_F \ell_{tr}} \hat{G}^{(2)}_{\boldsymbol{k},\boldsymbol{k}'} \ll \hat{G}^{(2)}_{\boldsymbol{k},\boldsymbol{k}'}. \tag{4.113}$$

As it only slightly renormalizes the second-order term, the crossed diagram can be neglected. Note that the other two fourth-order diagrams involve the integrals with a single Green function and are not small compared to $\hat{G}^{(2)}_{\boldsymbol{k},\boldsymbol{k}'}$.

To summarize, we see that all terms in the averaged series (4.98) are proportional to the Kronecker symbol $\delta_{\boldsymbol{k},\boldsymbol{k}'}$, allowing us to write the averaged Green function as

$$\langle\langle \hat{G}_{\boldsymbol{k},\boldsymbol{k}'} \rangle\rangle = \hat{G}_{\boldsymbol{k}} \delta_{\boldsymbol{k},\boldsymbol{k}'}. \tag{4.114}$$

The function $\hat{G}_{\boldsymbol{k}}$ is represented by an infinite series of *non-crossing* diagrams. This series sums up into a closed equation for $\hat{G}_{\boldsymbol{k}}$,

$$\hat{G}_{\boldsymbol{k}} = \hat{G}_{0\boldsymbol{k}} + \hat{G}_{0\boldsymbol{k}} \zeta \int \frac{d\boldsymbol{k}_1}{(2\pi)^2} \hat{G}_{\boldsymbol{k}_1} \hat{G}_{\boldsymbol{k}}, \tag{4.115}$$

By iteration we can check that Eq. (4.115) generates all possible non-crossing diagrams, for example, Eqs. (4.100), (4.104), and (4.105), as depicted in Fig. 4.13. The nonperturbative approximate method

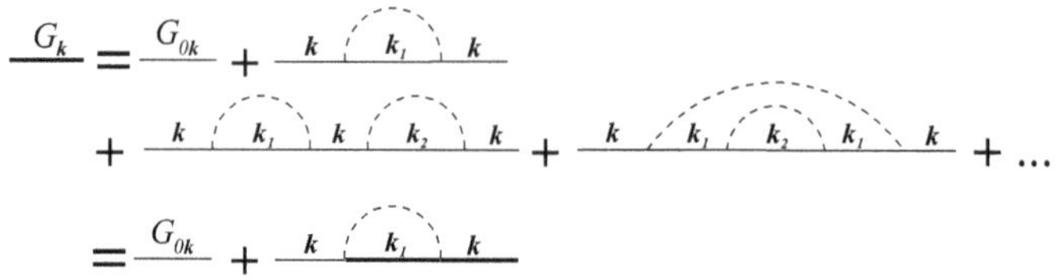

Figure 4.13 Diagrammatic representation of Eq. (4.115) for the averaged Green function $\hat{G}_{\boldsymbol{k}}$ (thick lines) in the self-consistent Born approximation.

employed above to derive Eq. (4.115) is known as the self-consistent Born approximation.

In the following we outline the solution procedure for Eq. (4.115). We seek the solution in the form

$$\hat{G}_{\boldsymbol{k}} = \frac{\hat{P}_{\boldsymbol{n}}}{E - \xi_{\boldsymbol{k}} - \Sigma_{\boldsymbol{n}}}. \tag{4.116}$$

Here, $\hat{P}_{\boldsymbol{n}}$ is the same projector matrix as in Eq. (4.82) for the unperturbed Green function. $\Sigma_{\boldsymbol{n}}$ is the unknown function called *self-energy*. It can depend only on the momentum direction $\boldsymbol{n}$ because $|\boldsymbol{k}|$ must be close to k_F. Inserting Eqs. (4.82) and (4.116) into Eq. (4.115) and using the identity

$$\hat{P}_{\boldsymbol{n}}\hat{P}_{\boldsymbol{n}_1}\hat{P}_{\boldsymbol{n}} = \hat{P}_{\boldsymbol{n}}\frac{1 + \mathbf{e}_{\boldsymbol{n}} \cdot \mathbf{e}_{\boldsymbol{n}_1}}{2}, \tag{4.117}$$

we obtain the equation for the self-energy:

$$\Sigma_{\boldsymbol{n}} = \zeta \int \frac{1}{E - \xi_{\boldsymbol{k}_1} - \Sigma_{\boldsymbol{n}_1}} \frac{1 + \mathbf{e}_{\boldsymbol{n}} \cdot \mathbf{e}_{\boldsymbol{n}_1}}{2} \frac{d\boldsymbol{k}_1}{(2\pi)^2}. \tag{4.118}$$

We again separate the angle averaging [see Eq. (4.109)] and transform the integral over $|\boldsymbol{k}_1|$ into integration over $\xi_{\boldsymbol{k}_1}$ with the limits from $-E_F$ to ∞. Under condition (4.81) the main contribution to the integral comes from the vicinity of the Fermi surface, $\xi_{\boldsymbol{k}_1} \approx 0$. We therefore, can extend the lower limit from $-E_F$ to $-\infty$,

$$\Sigma_{\boldsymbol{n}} = \zeta \int_0^{2\pi} \frac{d\phi_{\boldsymbol{n}_1}}{2\pi} \frac{1 + \mathbf{e}_{\boldsymbol{n}} \cdot \mathbf{e}_{\boldsymbol{n}_1}}{2} N \int_{-\infty}^{\infty} d\xi \frac{1}{E - \xi - \Sigma_{\boldsymbol{n}_1}}. \tag{4.119}$$

Moreover, the real part of $\Sigma_{\boldsymbol{n}_1}$ can be absorbed into the integration over ξ, leading to a purely imaginary self-energy

$$\Sigma_{\boldsymbol{n}} = \mp i|\Im\Sigma_{\boldsymbol{n}}|, \tag{4.120}$$

$$|\Im\Sigma_{\boldsymbol{n}}| = \zeta N \int_0^{2\pi} \frac{d\phi_{\boldsymbol{n}_1}}{2\pi} \frac{1 + \mathbf{e}_{\boldsymbol{n}} \cdot \mathbf{e}_{\boldsymbol{n}_1}}{2} \int_{-\infty}^{\infty} d\xi \frac{|\Im\Sigma_{\boldsymbol{n}_1}|}{(E-\xi)^2 + |\Im\Sigma_{\boldsymbol{n}_1}|^2}, \tag{4.121}$$

where $\Im$ denotes the imaginary part and $\mp$ are for the retarded and advanced functions, respectively. The ξ integral equals to π, so that with Eq. (4.92) for ζ we have

$$|\Im\Sigma_{\boldsymbol{n}}| = \frac{\hbar}{\tau_0} \int_0^{2\pi} \frac{d\phi_{\boldsymbol{n}_1}}{2\pi} \frac{1 + \mathbf{e}_{\boldsymbol{n}} \cdot \mathbf{e}_{\boldsymbol{n}_1}}{2}. \tag{4.122}$$

Inserting Eqs. (4.120) and (4.122) back into Eq. (4.116) and re-installing the superscripts R/A, we arrive at the final expressions for the disorder-averaged retarded and advanced Green functions,

$$\hat{G}_{\boldsymbol{k}}^{R/A} = \frac{\hat{P}_{\boldsymbol{n}}}{E_{R/A} - \xi_{\boldsymbol{k}}}, \qquad E_{R/A} = E \pm \frac{i\hbar}{2\tau}. \tag{4.123}$$

Thus, the disorder averaging introduces a finite single-particle life-time, τ, which is given by

$$\frac{1}{\tau} = \frac{2}{\hbar}|\Im\Sigma_{\boldsymbol{n}}| = \frac{1}{\tau_0} \int_0^{2\pi} \frac{d\phi_{\mathbf{n}_1}}{2\pi} (1 + \mathbf{e}_{\boldsymbol{n}} \cdot \mathbf{e}_{\boldsymbol{n}_1}), \tag{4.124}$$

where the unit vectors $\mathbf{n}$ and $\mathbf{n}_1$ specify the directions of the incident and scattered momentum states, respectively, and the unit vector $\mathbf{e}_{\boldsymbol{n}}$ can be written as

$$\mathbf{e}_{\boldsymbol{n}} = \mathbf{e}_{\parallel} + \mathbf{e}_{\perp}, \quad \mathbf{e}_{\parallel} = \boldsymbol{n}\sqrt{1 - e_{\perp}^2}, \quad \mathbf{e}_{\perp} = \frac{\mathcal{M}_{k_F}\, \mathbf{z}}{\sqrt{\mathcal{A}^2 k_F^2 + \mathcal{M}_{k_F}^2}}, \tag{4.125}$$

with $e_{\parallel} \equiv |\mathbf{e}_{\parallel}|$ and $e_{\perp} \equiv |\mathbf{e}_{\perp}|$. The angle averaging in Eq. (4.124) yields

$$\tau = \frac{\tau_0}{1 + e_{\perp}^2} = \frac{\tau_0}{1 + \mathcal{M}_{k_F}^2 / (\mathcal{A}^2 k_F^2 + \mathcal{M}_{k_F}^2)}. \tag{4.126}$$

This result is interesting because the time τ is shorter than the disorder-related time scale, τ_0 [see Eq. (4.92)], and involves the band structure parameters, the "mass" term $\mathcal{M}_{k_F}$ and constant $\mathcal{A}$. In order to understand this, we recall that $\mathcal{M}_{k_F}$ effectively acts as a Zeeman field, polarizing the carrier spin out of plane. This enables backscattering into an opposite momentum state with $\mathbf{n}_1 = -\mathbf{n}$ [see Eq. (4.124) and Fig. 4.8], otherwise forbidden for massless Dirac fermions [136]. Thus, the parameter $e_{\perp}^2$ in Eq. (4.126) quantifies the strength of the backscattering.

At this stage it is also convenient to introduce a related transport characteristic, the transport scattering time τ_{tr} [47], that differs from τ (4.124) by an extra factor $1 - \boldsymbol{n} \cdot \boldsymbol{n}_1$ under the integral

$$\frac{1}{\tau_{tr}} = \frac{1}{\tau_0} \int_0^{2\pi} \frac{d\phi_{\mathbf{n}_1}}{2\pi} (1 - \boldsymbol{n} \cdot \boldsymbol{n}_1)(1 + \mathbf{e}_{\boldsymbol{n}} \cdot \mathbf{e}_{\boldsymbol{n}_1}) = \frac{1 + 3e_\perp^2}{2\tau_0}. \quad (4.127)$$

This time determines the diffusion coefficient $D = v^2 \tau_{tr}/2$ and the classical conductivity via the Einstein relation,

$$\sigma_c = e^2 N D = \frac{e^2}{h} k_F v \tau_{tr} = \frac{e^2}{h} k_F \ell_{tr}. \quad (4.128)$$

The reason why τ_{tr}, and not τ, enters the conductivity is that the factor $1 - \boldsymbol{n} \cdot \boldsymbol{n}_1$ accounts for the non-equilibrium distribution of the carriers around the direction of the applied electric field. This effect is not captured in the life-time τ (4.124). In Section 4.4.7, the calculation of the classical conductivity is explained in more detail.

4.4.4 *Cooperon*

We now turn to averaging the direct product of the retarded and advanced Green functions [see Eq. (4.97)]. This reduces to calculating a specific two-particle Green function $C_{\alpha\beta\alpha'\beta'}(\boldsymbol{k}, \boldsymbol{k}', \omega)$, known as the Cooperon. Similar to the single-particle Green function, the Cooperon can be expressed by an infinite series of diagrams,

$$C_{\alpha\beta\alpha'\beta'}(\boldsymbol{k}, \boldsymbol{k}', \omega) = C^{(2)}_{\alpha\beta\alpha'\beta'}(\boldsymbol{k}, \boldsymbol{k}', \omega) + C^{(4)}_{\alpha\beta\alpha'\beta'}(\boldsymbol{k}, \boldsymbol{k}', \omega) + .., \quad (4.129)$$

as depicted in Fig. 4.14. As the crossed diagrams can again be omitted, the disorder averaging amounts to connecting the retarded and advanced functions by vertical dashed lines. This generates a series of "ladder" diagrams in which the retarded and advanced functions are simultaneously "dressed" according to Fig. 4.13. Since the zeroth order term

$$\langle\langle \hat{G}^R_{\boldsymbol{k},\boldsymbol{k}'}(E + \hbar\omega)\rangle\rangle \otimes \langle\langle \hat{G}^A_{\boldsymbol{k}',\boldsymbol{k}}\rangle\rangle$$

involves the independently averaged functions, we have extracted it, which makes the subsequent calculation of the conductivity correction easier. The ladder diagrams sum up into a closed equation for $C_{\alpha\beta\alpha'\beta'}(\boldsymbol{k}, \boldsymbol{k}', \omega)$, known as the Bethe–Salpeter equation:

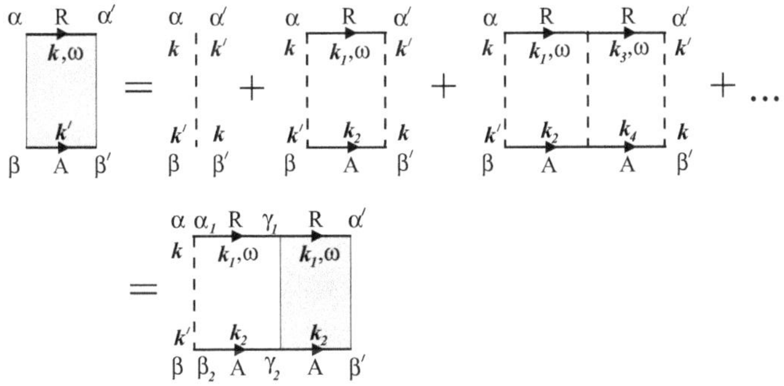

Figure 4.14 Bethe–Salpeter equation for Cooperon (shaded area). Vertical dashed lines represent the disorder correlator, for example, $\langle\langle V_{\alpha\alpha_1}(\mathbf{k}, \mathbf{k}_1) V_{\beta\beta_2}(\mathbf{k}', \mathbf{k}_2)\rangle\rangle$. Horizontal lines R and A denote dressed retarded and advanced Green functions, for example, $G^R_{\alpha_1,\gamma_1}(\mathbf{k}_1, E + \hbar\omega)$ and $G^A_{\beta_2,\gamma_2}(\mathbf{k}_2, E)$

$$\begin{aligned}C_{\alpha\beta\alpha'\beta'}(\mathbf{k}, \mathbf{k}', \omega) &= \langle\langle V_{\alpha\alpha'}(\mathbf{k}, \mathbf{k}') V_{\beta\beta'}(\mathbf{k}', \mathbf{k})\rangle\rangle \\ &+ \sum_{\alpha_1,\beta_2,\mathbf{k}_1,\mathbf{k}_2} \langle\langle V_{\alpha\alpha_1}(\mathbf{k}, \mathbf{k}_1) V_{\beta\beta_2}(\mathbf{k}', \mathbf{k}_2)\rangle\rangle \\ &\times G^R_{\alpha_1,\gamma_1}(\mathbf{k}_1, E + \hbar\omega) G^A_{\beta_2,\gamma_2}(\mathbf{k}_2, E) C_{\gamma_1\gamma_2,\alpha',\beta'}(\mathbf{k}_1, \mathbf{k}_2, \omega).\end{aligned} \tag{4.130}$$

With the correlation function (4.93) the above equation reads

$$\begin{aligned}C_{\alpha\beta\alpha'\beta'}(\mathbf{k}, \mathbf{k}', \omega) &= \frac{\zeta}{a}\delta_{\alpha\alpha'}\delta_{\beta\beta'} \\ &+ \zeta \int \frac{d\mathbf{k}_1}{(2\pi)^2} G^R_{\alpha,\gamma_1}(\mathbf{k}_1, E + \hbar\omega) G^A_{\beta,\gamma_2}(\mathbf{k} + \mathbf{k}' - \mathbf{k}_1, E) \\ &\times C_{\gamma_1\gamma_2,\alpha',\beta'}(\mathbf{k}_1, \mathbf{k} + \mathbf{k}' - \mathbf{k}_1, \omega).\end{aligned} \tag{4.131}$$

We note that the right-hand side is a function of the wave-vector sum $\mathbf{k} + \mathbf{k}'$. Therefore, the Cooperon depends on the *sum* of its wave-vector arguments:

$$C_{\alpha\beta\alpha'\beta'}(\mathbf{k}, \mathbf{k}', \omega) = C_{\alpha\beta\alpha'\beta'}(\mathbf{q}, \omega), \qquad \mathbf{q} = \mathbf{k} + \mathbf{k}'. \tag{4.132}$$

For $\mathbf{q} \to 0$ the wave vectors $\mathbf{k}$ and $\mathbf{k}'$ are antiparallel to each other. This is reminiscent of a Cooper pair, hence the name "Cooperon". In the weak localization theory the Cooperon (4.132) with $\mathbf{q} \to 0$ describes quantum interference of two states with the antiparallel momenta in a disordered medium.

It is convenient to change the normalization of the Cooperon,

$$C_{\alpha\beta\alpha'\beta'}(\boldsymbol{q},\omega) \to \frac{\zeta}{a}\frac{\tau_0}{\tau^2}C_{\alpha\beta\alpha'\beta'}(\boldsymbol{q},\omega), \tag{4.133}$$

and write Eq. (4.131) as

$$C_{\alpha\beta\alpha'\beta'}(\boldsymbol{q},\omega) = \frac{\tau^2}{\tau_0}\delta_{\alpha\alpha'}\delta_{\beta\beta'} + \zeta\int\frac{d\boldsymbol{k}}{(2\pi)^2} \times \sum_{\gamma'\delta'} G^R_{\alpha\gamma'}(\boldsymbol{k}, E+\hbar\omega)G^A_{\beta\delta'}(\boldsymbol{q}-\boldsymbol{k}, E)C_{\gamma'\delta'\alpha'\beta'}(\boldsymbol{q},\omega). \tag{4.134}$$

The standard way to solve this equation is to expand the Cooperon in the orthonormal basis of the two-electron spin states,

$$C_{\alpha\beta\alpha'\beta'}(\boldsymbol{q},\omega) = \sum_{ij} C^{ij}(\boldsymbol{q},\omega)\,\Psi^i_{\alpha\beta}\Psi^{j*}_{\alpha'\beta'}, \tag{4.135}$$

where the basis functions, $\Psi^i_{\alpha\beta}$, can be chosen as follows

$$\Psi^j = \frac{\sigma_j\sigma_y}{\sqrt{2}}, \qquad j = 0, x, y, z, \qquad \sum_{\alpha\beta}\Psi^j_{\alpha\beta}\Psi^{i*}_{\alpha\beta} = \delta_{ji}. \tag{4.136}$$

The index $j = 0$ labels the spin-singlet state, while $j = x, y,$ and z correspond to the three spin-triplet states. Consequently, there is one singlet Cooperon with the amplitude $C^{00}(\boldsymbol{q},\omega)$ and 15 triplet Cooperons. The next step is to derive equations for the Cooperon amplitudes $C^{ij}(\boldsymbol{q},\omega)$. This part of calculations is relegated to Appendix A.1. Here, we quote the result:

$$C^{ij}(\boldsymbol{q},\omega) = \frac{\tau^2}{\tau_0}\delta_{ij} + \frac{\tau}{\tau_0}\sum_{s=0,x,y,z}\mathrm{Tr}\langle\hat{P}_{-\boldsymbol{n}}\sigma_i\hat{P}_{\boldsymbol{n}}\sigma_s\rangle C^{sj}(\boldsymbol{q},\omega), \tag{4.137}$$

where $\hat{P}_{\boldsymbol{n}}$ is the spin matrix defined in Eq. (4.83), and brackets $\langle\ldots\rangle$ stand for the integral

$$\langle\ldots\rangle = \int_0^{2\pi}\frac{d\phi_{\boldsymbol{n}}}{2\pi}\ldots\times(1 - i\tau\omega + i\tau v\boldsymbol{n}\cdot\boldsymbol{q})^{-1}, \tag{4.138}$$

taken over the momentum directions on the Fermi surface.

4.4.5 *Cooperon for a Conventional Metal*

Since we have not used any concrete matrix $\hat{P}_{\boldsymbol{n}}$, Eq. (4.137) is valid also for a conventional metal with a spin-degenerate quadratic spectrum. In this case, the carrier Green function is given by Eq. (4.82) where $\hat{P}_{\boldsymbol{n}}$ should be replaced by the unit spin matrix, σ_0. Consequently, for the elastic scattering time τ one finds $\tau = \tau_0/2$. Inserting

$$\hat{P}_{\boldsymbol{n}} = \sigma_0 \qquad \text{and} \qquad \tau = \frac{\tau_0}{2} \tag{4.139}$$

into Eq. (4.137) and using $\mathrm{Tr}(\sigma_i\sigma_s) = 2\delta_{is}$, we find the Cooperon amplitudes for a conventional metal:

$$C^{ij}(\boldsymbol{q}, \omega) = \frac{\tau/2}{1 - \langle 1 \rangle}\delta_{ij}, \tag{4.140}$$

where $\langle 1 \rangle$ is the angle integral

$$\begin{aligned}\langle 1 \rangle &= \int_0^{2\pi} \frac{d\phi_{\boldsymbol{n}}}{2\pi}[1 - i\tau\omega + i\tau v\boldsymbol{n}\cdot\boldsymbol{q}]^{-1} \\ &= \int_0^{2\pi} \frac{d\phi_{\boldsymbol{n}}}{2\pi}[1 + (i\tau\omega - i\tau v\boldsymbol{n}\cdot\boldsymbol{q}) + (i\tau\omega - i\tau v\boldsymbol{n}\cdot\boldsymbol{q})^2 + \ldots].\end{aligned} \tag{4.141}$$

In the diffusive case defined by the conditions

$$\tau v|\boldsymbol{n}\cdot\boldsymbol{q}| \ll 1, \qquad \tau\omega \ll 1, \tag{4.142}$$

it is sufficient to expand Eq. (4.141) up to the second order, which after angle averaging yields

$$\langle 1 \rangle \approx 1 + i\tau\omega - \frac{\tau^2 v^2 \boldsymbol{q}^2}{2}. \tag{4.143}$$

Inserting this into Eq. (4.140), we have

$$C^{ij}(\boldsymbol{q}, \omega) = \frac{1/2}{D_0\boldsymbol{q}^2 - i\omega}\delta_{ij}, \qquad D_0 = \frac{v^2\tau}{2}, \tag{4.144}$$

where D_0 is the diffusion coefficient for a conventional metal, and the factor of 1/2 reflects the chosen normalization [see Eq. (4.133)].

A few comments about Eq. (4.144) are in order. First, the matrix of the Cooperon amplitudes C^{ij} is diagonal. Second, the Cooperons are divergent for $\boldsymbol{q}, \omega \to 0$. This eventually leads to the logarithmic divergence of the WL correction in Eq. (4.51). Third, the AC electric field produces dephasing $\hbar\omega$ in energies of the counter-propagating states. The dephasing is accounted for by the imaginary term $-i\omega$ in the Cooperon denominator. If the dephasing is dominated by other sources, typically by electron–electron or electron–phonon interactions, then $-i\omega$ is replaced by the dephasing rate $\tau_\phi^{-1}(T)$. It is a function of temperature depending on the type of the interaction.

4.4.6 *Cooperon for a Helical Metal*

For a helical metal $\hat{P}_{\boldsymbol{n}}$ is a projector matrix defined in Eq. (4.83). Inserting $\hat{P}_{\boldsymbol{n}}$ from Eq. (4.83) into Eq. (4.137) and evaluating the traces of the Pauli matrices we find

$$C^{0j} = \frac{\tau^2}{\tau_0}\delta_{0j} + \frac{\tau}{2\tau_0}\langle 1 - \mathbf{e}_+ \cdot \mathbf{e}_- \rangle C^{0j} + \frac{\tau}{2\tau_0}\sum_{b=x,y,z}\langle (\mathbf{e}_+ - \mathbf{e}_- + i\mathbf{e}_+ \times \mathbf{e}_-)\cdot \mathbf{b}\rangle C^{bj}, \tag{4.145}$$

$$\begin{aligned} C^{aj} = \frac{\tau^2}{\tau_0}\delta_{aj} &+ \frac{\tau}{2\tau_0}\langle (\mathbf{e}_+ - \mathbf{e}_- - i\mathbf{e}_+ \times \mathbf{e}_-)\cdot \mathbf{a}\rangle C^{0j} \\ &+ \frac{\tau}{2\tau_0}\sum_{b=x,y,z}\langle (1 + \mathbf{e}_+ \cdot \mathbf{e}_-)\mathbf{a}\cdot\mathbf{b} - i(\mathbf{e}_+ + \mathbf{e}_-)\cdot \mathbf{a}\times\mathbf{b} \\ &- (\mathbf{e}_+ \cdot \mathbf{a})(\mathbf{e}_- \cdot \mathbf{b}) - (\mathbf{e}_- \cdot \mathbf{a})(\mathbf{e}_+ \cdot \mathbf{b})\rangle C^{bj}. \end{aligned} \tag{4.146}$$

We separated the amplitudes C^{0j} and $C^{(a,b)j}$, where a and b both run over the triplet indices x, y, and z only. Respectively, vectors $\mathbf{a}$ and $\mathbf{b}$ run over the unit vector basis of the Cartesian system. We also introduced a shorthand notation for the unit vectors

$$\mathbf{e}_\pm = \mathbf{e}_{\pm\boldsymbol{n}}. \tag{4.147}$$

Equations (4.145) and (4.146) are the central result of this section. Unlike the conventional metal [see Eq. (4.140)], the Cooperon matrix C^{ij} is nondiagonal, and the singlet and triplet Cooperons are coupled to each other in Eqs. (4.145) and (4.146). This is the consequence of the spin–momentum locking of helical states. Although our starting point was the TI thin film model, the applicability of Eqs. (4.145) and (4.146) is, in fact, much broader. For instance, they also describe a TI surface with the hexagonal warping. The warping anisotropy is encoded in the dependence of $\mathbf{e}_{\boldsymbol{n}}$ on the momentum direction $\boldsymbol{n}$. This case is discussed in the next section.

In the following, we solve Eqs. (4.145) and (4.146) for the particular case of TI thin films. Our goal is to understand how the carrier backscattering affects the Cooperons. As discussed in the preceding subsection, the backscattering strength is characterized by the out-of-plane component, $\mathbf{e}_\perp$, of the unit vectors (4.147). It is therefore convenient to separate the in- and out-of-plane components,

$$\mathbf{e}_\pm = \pm\mathbf{e}_\parallel + \mathbf{e}_\perp, \tag{4.148}$$

where $\mathbf{e}_\parallel$ and $\mathbf{e}_\perp$ are defined by Eq. (4.125). In view of the identities

$$\mathbf{e}_+ + \mathbf{e}_- = 2\mathbf{e}_\perp, \quad \mathbf{e}_+ - \mathbf{e}_- = 2\mathbf{e}_\parallel, \tag{4.149}$$

$$\mathbf{e}_+ \cdot \mathbf{e}_- = 1 - 2\mathbf{e}_\parallel^2, \quad \mathbf{e}_+ \times \mathbf{e}_- = 2\mathbf{e}_\parallel \times \mathbf{e}_\perp, \tag{4.150}$$

Eqs. (4.145) and (4.146) reduce to

$$C^{0j} = \frac{\tau^2}{\tau_0}\delta_{0j} + \frac{\tau}{\tau_0}\langle \mathbf{e}_\parallel^2\rangle C^{0j} + \frac{\tau}{\tau_0}\sum_{b=x,y,z}\langle(\mathbf{e}_\parallel + i\mathbf{e}_\parallel \times \mathbf{e}_\perp)\cdot\mathbf{b}\rangle C^{bj}, \tag{4.151}$$

$$C^{aj} = \frac{\tau^2}{\tau_0}\delta_{aj} + \frac{\tau}{\tau_0}\langle(\mathbf{e}_\parallel - i\mathbf{e}_\parallel \times \mathbf{e}_\perp)\cdot\mathbf{a}\rangle C^{0j} + \frac{\tau}{\tau_0}\sum_{b=x,y,z}\langle \mathbf{e}_\perp^2(\mathbf{a}\cdot\mathbf{b}) -$$
$$i\mathbf{e}_\perp\cdot(\mathbf{a}\times\mathbf{b}) + (\mathbf{e}_\parallel\cdot\mathbf{a})(\mathbf{e}_\parallel\cdot\mathbf{b}) - (\mathbf{e}_\perp\cdot\mathbf{a})(\mathbf{e}_\perp\cdot\mathbf{b})\rangle C^{bj}. \tag{4.152}$$

Explicitly, these equations are

$$\left[\frac{\tau_0}{\tau} - \langle e_\parallel^2\rangle\right] C^{0j} - \langle e_x + ie_ye_\perp\rangle C^{xj} - \langle e_y - ie_xe_\perp\rangle C^{yj} = \tau\delta_{0j}, \tag{4.153}$$

$$-\langle e_x - ie_ye_\perp\rangle C^{0j} + \left[\frac{\tau_0}{\tau} - \langle e_\perp^2 + e_x^2\rangle\right] C^{xj} - \langle e_xe_y - ie_\perp\rangle C^{yj} = \tau\delta_{xj}, \tag{4.154}$$

$$-\langle e_y + ie_xe_\perp\rangle C^{0j} - \langle e_ye_x + ie_\perp\rangle C^{xj} + \left[\frac{\tau_0}{\tau} - \langle e_\perp^2 + e_y^2\rangle\right] C^{yj} = \tau\delta_{yj}, \tag{4.155}$$

$$C^{zj} = \frac{\tau^2}{\tau_0}\delta_{zj}. \tag{4.156}$$

The decoupled Cooperons C^{zj} are momentum- and frequency-independent and can be safely neglected. To calculate the conductivity correction we need the diagonal Cooperons $C^{00}(\boldsymbol{q}, \omega)$, $C^{xx}(\boldsymbol{q}, \omega)$, and $C^{yy}(\boldsymbol{q}, \omega)$. These are obtained from the coupled Eqs. (4.153)–(4.155) in which index j should be subsequently set to 0, x, and y.

The coefficients in Eqs. (4.153)–(4.155) are evaluated by expanding the denominator in Eq. (4.138) in powers of the Cooperon wave vector, $\boldsymbol{q}$, and frequency, ω, under condition (4.142). In doing so we keep the lowest order terms that yield the nonzero angle average $\langle \ldots \rangle$ in Eqs. (4.153)–(4.155). Under these conditions one obtain the following expressions for the diagonal Cooperons:

$$C^{00}(\boldsymbol{q},\omega) = \frac{1}{D\boldsymbol{q}^2 + \tau_{\mathcal{M}}^{-1} - i\omega}, \qquad \tau_{\mathcal{M}}^{-1} = \frac{2e_\perp^2}{\tau}, \tag{4.157}$$

$$C^{xx}(\boldsymbol{q},\omega) = \frac{2\tau}{e_\parallel^2} \frac{2 - e_\parallel^4 \cos^2\phi_{\boldsymbol{q}}}{2 + 5e_\perp^2 + e_\perp^4} + \frac{2\tau(\tau_{\mathcal{M}}^{-1} - i\omega)}{e_\parallel^4} \frac{(1 + 3e_\perp^2)^2 - e_\parallel^6 \sin^2\phi_{\boldsymbol{q}}}{2 + 5e_\perp^2 + e_\perp^4} C^{00}(\boldsymbol{q},\omega), \tag{4.158}$$

$$C^{yy}(\boldsymbol{q},\omega) = \frac{2\tau}{e_\parallel^2} \frac{2 - e_\parallel^4 \sin^2\phi_{\boldsymbol{q}}}{2 + 5e_\perp^2 + e_\perp^4} + \frac{2\tau(\tau_{\mathcal{M}}^{-1} - i\omega)}{e_\parallel^4} \frac{(1 + 3e_\perp^2)^2 - e_\parallel^6 \cos^2\phi_{\boldsymbol{q}}}{2 + 5e_\perp^2 + e_\perp^4} C^{00}(\boldsymbol{q},\omega). \tag{4.159}$$

In Eq. (4.157) $D = D_0(2 + 5e_\perp^2 + e_\perp^4)/e_\parallel^2$ is the diffusion coefficient with the prefactor defined in Eq. (4.144). In Eqs. (4.158) and (4.159) $\phi_{\boldsymbol{q}}$ is the angle indicating the direction of the Cooperon momentum: $\boldsymbol{q} = |\boldsymbol{q}| \cdot (\cos\phi_{\boldsymbol{q}}, \sin\phi_{\boldsymbol{q}}, 0)$.

Now, we can see how the backscattering influences the Cooperons in Eqs. (4.157)–(4.159). There are two main effects. First, it introduces a new time scale, $\tau_{\mathcal{M}}$, that determines the rate of backscattering $\tau_{\mathcal{M}}^{-1}$. The latter enters the singlet Cooperon C^{00} in the same fashion as the dephasing rate $-i\omega$. The backscattering eliminates the divergence of C^{00} at $\boldsymbol{q}, \omega \to 0$, meaning that at time scales larger than $\tau_{\mathcal{M}}$ the quantum interference is suppressed. In order to be observable the time $\tau_{\mathcal{M}}$ itself must be larger than the elastic life time:

$$\frac{\tau}{\tau_{\mathcal{M}}} = 2e_\perp^2 \ll 1, \tag{4.160}$$

which requires small values of the parameter $e_\perp^2$. Under this condition the diffusion coefficient is

$$D \approx 2D_0. \tag{4.161}$$

The factor of 2 reflects the specifics of the Dirac-like spectrum [135, 136, 138]. Second, the backscattering modifies the triplet Cooperons C^{xx} and C^{yy} in Eqs. (4.158) and (4.159). Despite their complicated structure, Eqs. (4.158) and (4.159) are easy to interpret. The first terms never become divergent and can be safely neglected. The second terms are proportional to $C^{00}(\boldsymbol{q}, \omega)$ with the prefactors depending on the backscattering rate $\tau_{\mathcal{M}}^{-1}$ and $-i\omega$. For $\omega \to 0$ they scale with the small parameter $e_{\perp}^{2}$. It can be shown that the backscattering also generates other triplet Cooperons, C^{xy} and C^{yx}, which are proportional to $e_{\perp}^{3}$.

4.4.7 *Quantum Conductivity Correction*

With the averaged Green functions $\hat{G}_{\boldsymbol{k}}^{R,A}$ and the Cooperon $C_{\alpha\beta\alpha'\beta'}(\boldsymbol{q}, \omega)$ at hand, we proceed to calculate the conductivity from Kubo formula (4.59). Our goal is to express the disorder-averaged conductivity in terms of the known $\hat{G}_{\boldsymbol{k}}^{R,A}$ and $C_{\alpha\beta\alpha'\beta'}(\boldsymbol{q}, \omega)$. Inserting series expansion (4.65)–(4.68) for the exact Green functions into Kubo formula (4.59) and averaging it over disorder configurations, we obtain an infinite series:

$$\langle\langle\sigma_{xx}\rangle\rangle = \sigma_{xx}^{(0)} + \sigma_{xx}^{(2)} + \sigma_{xx}^{(4)} + \ldots, \qquad (4.162)$$

where the superscript indicates the order in the scattering potential. The terms up to the fourth order are shown diagrammatically in Fig. 4.15. Since Kubo formula (4.59) involves the traces over the momentum and spin variables, each term can be pictured as a "bubble" formed by the retarded and advanced Green functions connecting the velocity operators $\hat{v}_{\boldsymbol{k}}^{x}$ in the direction around the loop (which is indicated by the arrows in Fig. 4.15).

In order to illustrate the summation procedure we begin with the second order diagrams. The first two of them are corrections to the bare retarded and advanced Green functions (*cf.* Fig. 4.13). These diagrams can be absorbed into the zeroth order term with the correspondingly modified retarded and advanced functions. The third diagram leads to the renormalization of one of the velocity operators. This is known as a vertex correction. Such corrections are conventionally represented by ladder diagrams in Fig. 4.16. Proceeding to the fourth order terms in Fig. 4.15, we see that the

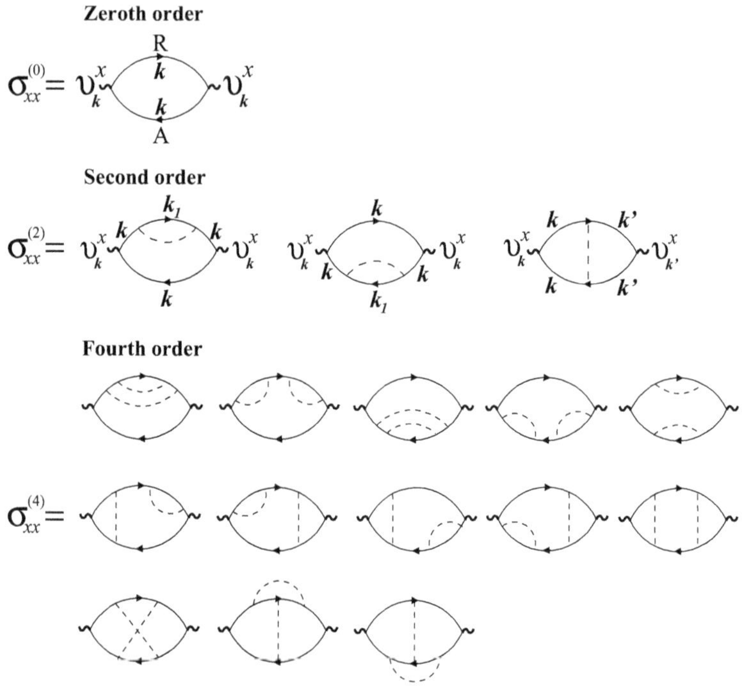

Figure 4.15 Diagrammatic representation of Kubo conductivity up to fourth order in disorder potential.

first two rows are corrections to the bare Green functions and/or the vertex corrections renormalizing $\hat{v}^x_{\boldsymbol{k}}$ according to Fig. 4.16. Again, these can be absorbed into the zeroth-order term by simultaneously renormalizing the Green functions and one of the velocity operators. Similar corrections appear in all orders and can be summed up according to the aforedescribed recipe. This summation scheme leads to the classical conductivity of a disordered metal, σ_c, which is depicted in Fig. 4.17. The corresponding analytical expression is

$$\sigma_c = \frac{e^2}{2\pi\omega}\int dE[f(E) - f(E+\hbar\omega)]\int\frac{d\boldsymbol{k}}{(2\pi)^2}\mathrm{Tr}[\hat{v}^x_{\boldsymbol{k}}\hat{G}^R_{\boldsymbol{k},E+\hbar\omega}\hat{\mathcal{V}}^x_{\boldsymbol{k}}\hat{G}^A_{\boldsymbol{k},E}], \tag{4.163}$$

where $\hat{G}^{R,A}_{\boldsymbol{k}}$ are the averaged Green functions given by Eq. (4.123), and $\hat{\mathcal{V}}^x_{\boldsymbol{k}}$ denotes the velocity operator renormalized by the ladder diagrams in Fig. 4.16. These diagrams sum up into the equation for $\hat{\mathcal{V}}^x_{\boldsymbol{k}}$:

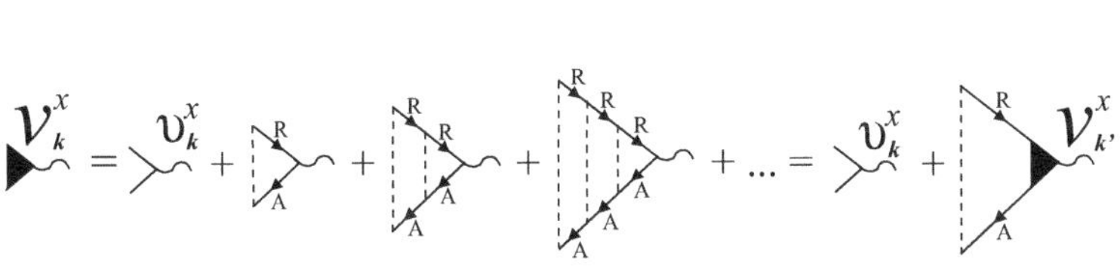

Figure 4.16 Ladder diagrams for vertex corrections to bare velocity operator $\hat{v}_{\boldsymbol{k}}^{x}$. Black triangle denotes the renormalized operator, $\hat{\mathcal{V}}_{\boldsymbol{k}}^{x}$.

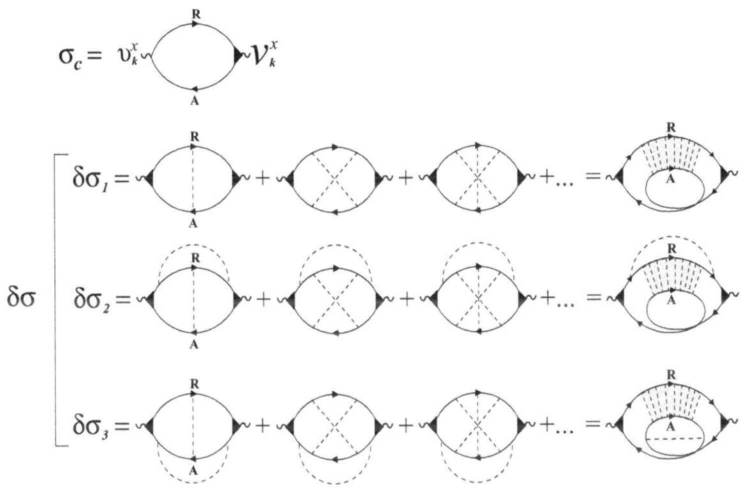

Figure 4.17 Diagrammatic representation of the classical conductivity, σ_c, and the quantum correction, $\delta\sigma$. Shaded area denotes the Cooperon.

$$\hat{\mathcal{V}}_{\boldsymbol{k}}^{x} = \hat{v}_{\boldsymbol{k}}^{x} + \zeta \int \frac{d\boldsymbol{k}}{(2\pi)^2} \hat{G}_{\boldsymbol{k},E+\hbar\omega}^{R} \hat{\mathcal{V}}_{\boldsymbol{k}}^{x} \hat{G}_{\boldsymbol{k},E}^{A}. \tag{4.164}$$

It is possible to show that the solution for $\hat{\mathcal{V}}_{\boldsymbol{k}}^{x}$ differs from the bare velocity just by a prefactor depending on the ratio of the transport and elastic scattering times (Ref. [47]),

$$\hat{\mathcal{V}}_{\boldsymbol{k}}^{x} = \left(\frac{\tau_{tr}}{\tau}\right) \hat{v}_{\boldsymbol{k}}^{x}, \qquad \frac{\tau_{tr}}{\tau} = \frac{1+3e_{\perp}^{2}}{2(1+e_{\perp}^{2})}, \tag{4.165}$$

where τ and τ_{tr} were introduced earlier in Eqs. (4.124) and (4.127). After this, the classical conductivity in Eq. (4.128) can be readily calculated, yielding for $\omega\,\tau \ll 1$ the Einstein relation (4.128) introduced in Section 4.4.3.

We now turn to the last three fourth-order diagrams in Eq. (4.15). Clearly, these are neither Green function nor vertex corrections. They cannot be incorporated into σ_c by renormalizing $G_{\mathbf{k}}^{R/A}$ or $\hat{v}_{\mathbf{k}}^{x}$. Therefore, the last three diagrams in Eq. (4.15) belong to the *quantum correction* to the classical conductivity. More precisely, the sum of such diagrams from all orders represents the quantum correction, $\delta\sigma$, as depicted in Fig. 4.17. The relevant diagrams are those in which all the dashed lines connecting the retarded and advanced functions cross each other within the bubble. Such diagrams are related to the Cooperon. Indeed, by twisting the A line as shown in Fig. 4.17 we see that all the crossed lines sum up into the ladder diagram for the Cooperon (shaded area, *cf.* Fig. 4.14). Thus, the conductivity correction $\delta\sigma$ consists of three contributions,

$$\delta\sigma = \delta\sigma_1 + \delta\sigma_2 + \delta\sigma_3, \tag{4.166}$$

The diagrams for $\delta\sigma_1$, $\delta\sigma_2$, and $\delta\sigma_3$, which all incorporate the Cooperon, are also called the *Hikami boxes*. Since the diagrams for the last two corrections in Fig. 4.17 are similar, $\delta\sigma_3$ appears to be equal to $\delta\sigma_2$. This reduces the problem to calculating $\delta\sigma_1$ and $\delta\sigma_2$. The corresponding analytical expressions are

$$\begin{aligned}\delta\sigma_1 = & \frac{e^2\hbar}{\pi N\tau^2}\int\frac{dE}{2\pi\omega}[f(E)-f(E+\hbar\omega)]\int\frac{d\mathbf{q}}{(2\pi)^2}C_{\beta\beta'\gamma\gamma'}(\mathbf{q},\omega)\\ & \times\int\frac{d\mathbf{k}}{(2\pi)^2}(\hat{G}^{A}_{\mathbf{k},E}\,\hat{v}^{x}_{\mathbf{k}}\,\hat{G}^{R}_{\mathbf{k},E+\hbar\omega})_{\gamma'\beta}(\hat{G}^{R}_{\mathbf{q}-\mathbf{k},E+\hbar\omega}\,\hat{v}^{x}_{\mathbf{q}-\mathbf{k}}\,\hat{G}^{A}_{\mathbf{q}-\mathbf{k},E})_{\gamma\beta'},\end{aligned} \tag{4.167}$$

$$\begin{aligned}\delta\sigma_2 = & \frac{e^2\hbar^2}{\pi^2N^2\tau_0\tau^2}\int\frac{dE}{2\pi\omega}[f(E)-f(E+\hbar\omega)]\int\frac{d\mathbf{q}}{(2\pi)^2}C_{\beta\beta'\gamma\gamma'}(\mathbf{q},\omega)\\ & \times\int\frac{d\mathbf{k}}{(2\pi)^2}\int\frac{d\mathbf{k'}}{(2\pi)^2}(\hat{G}^{A}_{\mathbf{k},E}\,\hat{v}^{x}_{\mathbf{k}}\,\hat{G}^{R}_{\mathbf{k},E+\hbar\omega}\hat{G}^{R}_{\mathbf{q}-\mathbf{k'},E+\hbar\omega})_{\gamma'\beta}\\ & \times(\hat{G}^{R}_{\mathbf{q}-\mathbf{k},E+\hbar\omega}\,\hat{G}^{R}_{\mathbf{k'},E+\hbar\omega}\,\hat{v}^{x}_{\mathbf{k'}}\,\hat{G}^{A}_{\mathbf{k'},E})_{\gamma\beta'},\end{aligned} \tag{4.168}$$

where summation over the spin indices is assumed. Since in the $\mathbf{k}$ integrals only the vicinity of the Fermi surface matters, we change the integration variables: $\frac{d\mathbf{k}}{(2\pi)^2}\to\frac{d\phi_n}{2\pi}Nd\xi$, and integrate over ξ. The calculation procedure follows the same steps as in the case with the Cooperon equation (see Appendix A.2). The results are

$$\delta\sigma_1 = \frac{8e^2D_0}{hv^2}\overline{[\hat{P}_{\mathbf{n}}\hat{v}^{x}_{\mathbf{n}}\hat{P}_{\mathbf{n}}]_{\gamma'\beta}[\hat{P}_{-\mathbf{n}}\hat{v}^{x}_{-\mathbf{n}}\hat{P}_{-\mathbf{n}}]_{\gamma\beta'}}\int\frac{d\mathbf{q}}{(2\pi)^2}C_{\beta\beta'\gamma\gamma'}(\mathbf{q},\omega), \tag{4.169}$$

$$\delta\sigma_2 = -\frac{8e^2 D_0 \tau}{hv^2 \tau_0} \overline{[\hat{P}_{\boldsymbol{n}} \hat{\mathcal{V}}^x_{\boldsymbol{n}} \hat{P}_{\boldsymbol{n}}]_{\gamma'\beta_1} [\hat{P}_{-\boldsymbol{n}}]_{\gamma\gamma_1}} \\ \times \overline{[\hat{P}_{-\boldsymbol{n}'}]_{\beta_1\beta} [\hat{P}_{\boldsymbol{n}'} \hat{\mathcal{V}}^x_{\boldsymbol{n}'} \hat{P}_{\boldsymbol{n}'}]_{\gamma_1\beta'}} \int \frac{d\boldsymbol{q}}{(2\pi)^2} C_{\beta\beta'\gamma\gamma'}(\boldsymbol{q}, \omega), \tag{4.170}$$

where the bar, as usual, denotes the angle averaging.

4.4.8 *Conductivity Correction for a Conventional Metal*

Equations (4.169) and (4.170) recover the well known result, Eq. (4.51), for the WL conductivity correction for a conventional metal with the spin-degenerate quadratic dispersion. In this case, we use $\hat{P}_{\boldsymbol{n}} = \sigma_0$, $\hat{\mathcal{V}}^x_{\boldsymbol{n}} = \hat{v}^x_{\boldsymbol{n}} = v\, n_x \sigma_0$, and Eqs. (4.135) and (4.144) for the Cooperon. Inserting these into Eqs. (4.169) and (4.170) and doing the angle averaging, we have

$$\delta\sigma_1 = -\frac{4e^2 D_0}{h} \int \frac{d\boldsymbol{q}}{(2\pi)^2} [C^{xx} + C^{yy} + C^{zz} - C^{00}] \\ = -\frac{2e^2}{(2\pi)h} \int_0^{D_0 q_m^2} d(D_0 q^2) \frac{1}{D_0 q^2 - i\omega}, \tag{4.171}$$

$$\delta\sigma_2 = 0. \tag{4.172}$$

It is worth noting that the negative sign in front of Eq. (4.171) reflects the opposite directions of the velocities of the interfering states. The correction $\delta\sigma_2$ vanishes because the angle averaging in Eq. (4.170) is reduced to the product of the average velocities, each being zero. When evaluating the integral, we should keep in mind that the expression for the Cooperon is valid in the diffusive regime for small values of $\tau D_0 q^2$ [*cf.*, Eq. (4.142)]. Therefore, the integration must be cut when $\tau D_0 q^2$ becomes of order of 1, which yields the upper integration limit $D_0 q_m^2 = \tau^{-1}$. Replacing further $-i\omega \to \tau_\phi^{-1}$ for generality, we obtain Eq. (4.51) from Eq. (4.171).

4.4.9 *Conductivity Correction for a Helical Metal*

In the case of a helical metal, $\hat{P}_{\boldsymbol{n}}$ in Eqs. (4.169) and (4.170) is a projector matrix defined in Eq. (4.83). With the help of the identity

$$\hat{P}_{\boldsymbol{n}} \hat{\mathcal{V}}^x_{\boldsymbol{n}} \hat{P}_{\boldsymbol{n}} = \frac{\tau_{tr}}{\tau} v\, n_x \hat{P}_{\boldsymbol{n}}, \tag{4.173}$$

Eqs. (4.169) and (4.170) can be simplified to

$$\delta\sigma_1 = -\frac{8e^2 D_0}{h}\left(\frac{\tau_{tr}}{\tau}\right)^2 \overline{n_x^2[\hat{P}_{\boldsymbol{n}}]_{\gamma'\beta}[\hat{P}_{-\boldsymbol{n}}]_{\gamma\beta'}} \int \frac{d\boldsymbol{q}}{(2\pi)^2} C_{\beta\beta'\gamma\gamma'}(\boldsymbol{q}, \omega), \tag{4.174}$$

$$\delta\sigma_2 = -\frac{8e^2 D_0}{h}\frac{\tau_{tr}^2}{\tau\tau_0} \overline{n_x[\hat{P}_{\boldsymbol{n}}]_{\gamma'\beta_1}[\hat{P}_{-\boldsymbol{n}}]_{\gamma\gamma_1}} \tag{4.175}$$
$$\times \overline{n_x[\hat{P}_{-\boldsymbol{n}'}]_{\beta_1\beta}[\hat{P}_{\boldsymbol{n}'}]_{\gamma_1\beta'}} \int \frac{d\boldsymbol{q}}{(2\pi)^2} C_{\beta\beta'\gamma\gamma'}(\boldsymbol{q}, \omega).$$

Unlike the case of a conventional metal, $\hat{P}_{\boldsymbol{n}}$ contains the spin helicity $\boldsymbol{\sigma} \cdot \mathbf{e}_{\boldsymbol{n}}$, which contributes to the angle averaging, leading to a nonzero correction $\delta\sigma_2$. We proceed by expanding the Cooperon in the basis of the singlet and triplet states [see Eq. (4.135)] and performing the summations over the spin indices. After some algebra the conductivity corrections in Eqs. (4.174) and (4.175) can be expressed in terms of the singlet and triplet Cooperon amplitudes:

$$\delta\sigma_1 = -\frac{4e^2 D_0}{h}\left(\frac{\tau_{tr}}{\tau}\right)^2 \int \frac{d\boldsymbol{q}}{(2\pi)^2}\Big[-\overline{n_x^2}(1-e_\perp^2)C^{00} \tag{4.176}$$
$$+ \overline{n_x^2}(1-e_y^2)C^{xx} + \overline{n_x^2}(1-e_x^2)C^{yy} + \overline{n_x^2} ie_\perp(C^{xy} - C^{yx})\Big],$$

$$\delta\sigma_2 = \frac{2e^2 D_0}{h}\frac{\tau_{tr}^2}{\tau\tau_0} e_\parallel^2 (\overline{n_x^2})^2 \int \frac{d\boldsymbol{q}}{(2\pi)^2}\Big[-(1+e_\perp^2)C^{00}$$
$$+ C^{xx} + e_\perp^2 C^{yy} + ie_\perp(C^{xy} - C^{yx})\Big]. \tag{4.177}$$

Noticing further that on average over the directions of $\boldsymbol{q}$, the triplet Cooperons (4.158) and (4.159) coincide, we can write the net conductivity correction as

$$\delta\sigma = \delta\sigma_1 + 2\delta\sigma_2 = -\frac{2e^2 D_0}{h}\left(\frac{\tau_{tr}}{\tau}\right)^2 \int \frac{qdq}{2\pi}\Bigg[-\frac{1-e_\perp^2}{2}\overline{C^{00}}$$
$$+\frac{1+3e_\perp^2}{2}\overline{C^{xx}} + \frac{ie_\perp(1+3e_\perp^2)}{2(1+e_\perp^2)}(\overline{C^{xy}} - \overline{C^{xy}})\Bigg], \tag{4.178}$$

where the bar denotes the angle averaging $\overline{(\ldots)} = \int_0^{2\pi} \ldots d\phi_{\boldsymbol{q}}/2\pi$ with $\phi_{\boldsymbol{q}}$ defined in the text after Eq. (4.159). Note that, Eq. (4.178) has been obtained without the assumption that $e_\perp^2$ is small.

In Eq. (4.178), the singlet Cooperon C^{00} enters with the negative sign because in the singlet channel the spin projections of the interfering states are antiparallel. This can also be viewed as the manifestation of an intrinsic π Berry phase [135]. Given the other minus sign from the opposite velocities, the singlet contribution to $\delta\sigma$ is positive, which corresponds to the WAL effect. In contrast, in the triplet channels the interference occurs between the states with the parallel spin projections. The triplet contribution to $\delta\sigma$ is negative and proportional to $e_\perp^2$ characterizing the out-of-plane spin polarization. For sufficiently small $e_\perp^2 \ll 1$, it suffices to keep only the singlet Cooperon (4.157). Then, the q integral is calculated in the same manner as described in Section 4.4.8:

$$\delta\sigma \approx \frac{e^2}{2\pi h} 2 \int_0^{Dq_m^2} d(Dq^2) \frac{1/2}{Dq^2 + \tau_{\mathcal{M}}^{-1} - i\omega} \tag{4.179}$$

$$= \frac{e^2}{2\pi h} \ln \frac{\tau^{-1}}{\tau_{\mathcal{M}}^{-1} + \tau_\phi^{-1}}. \tag{4.180}$$

Here, we took $D \approx 2D_0$ and $\tau_{tr}/\tau \approx 2$ in the leading order in $e_\perp^2 \ll 1$ [see Eqs. (4.161) and (4.165), respectively]. The factor of 2 from the scattering times cancels with the factor of 1/2 that originates from the correction $\delta\sigma_2$. As a result, in the leading order in the small parameter $e_\perp^2$ the net correction in Eq. (4.180) has the standard universal prefactor. If we now add the contribution of the "−" subsystem in Eq. (4.41), we arrive at Eq. (4.58) for the WAL correction for a TI thin film.

4.5 Weak Antilocalization on TI Surfaces

As already mentioned, the surface states in Bi_2Se_3 or Bi_2Te_3 have a single-cone anisotropic spectrum [see also Eq. (4.35) and Fig. 4.3] described by Hamiltonian (4.34) with the hexagonal warping given by Eq. (4.32). Since the Hamiltonian is TRS invariant, it is quite natural to expect the WAL conductivity correction to be typical of the symplectic symmetry class of disordered systems. Here, we demonstrate this by straightforward calculation, assuming again that the Fermi level lies in the conduction band [see Figs. 4.3 and 4.11] and the band separation is large enough to enable the metallic transport regime [see Eqs. (4.46) and (4.48)]. The warping will be treated as a weak perturbation of the surface bands under condition

$$\frac{W^2k^4}{2\mathcal{A}^2} \ll 1, \tag{4.181}$$

which is justified for not very large surface Fermi energies. In this case the main effect of the warping on the spectrum is to increase the carrier energy $E(k, \phi_{\mathbf{n}})$ on average over all the angles $\phi_{\mathbf{n}}$ [see Eq. (4.35)]. This amounts to replacing $\mathcal{M}_{\mathbf{k}}^2$ by its angle average,

$$\mathcal{M}_{\mathbf{k}}^2 \Longrightarrow \overline{\mathcal{M}_{\mathbf{k}}^2} = \frac{W^2k^6}{2}, \tag{4.182}$$

in Eq. (4.35). In fact, the same replacement can be done in all *even* functions of $\mathcal{M}_{\mathbf{k}}$, for example, in the Fermi momentum, Fermi velocity, DOS, etc. This allows us to use Eq. (4.123) for the retarded and advanced Green functions, Cooperon equations (4.145) and (4.146) as well as the Hikami boxes (4.169) and (4.170) derived in the preceding section.

The angular dependence of the warping is accounted for in the carrier helicity $\boldsymbol{\sigma} \cdot \mathbf{e}_n$, where the vector $\mathbf{e}_n$ can be written as

$$\mathbf{e}_n = \mathbf{e}_{\parallel} + \mathbf{e}_{\perp}, \quad \mathbf{e}_{\parallel} \approx \boldsymbol{n}\sqrt{1 - \overline{e_{\perp}^2}}, \quad \mathbf{e}_{\perp} \approx \frac{\mathcal{M}_{k_F\,\boldsymbol{n}}}{\mathcal{A}k_F}\mathbf{z}. \tag{4.183}$$

The warping generates the vector component $\mathbf{e}_{\perp}$ that describes the out-of-plane angular modulation of the spin texture on the Fermi surface. This is the main nontrivial effect of the warping which needs to be taken into account in the calculations of the transport properties of the TI surface state.

4.5.1 *Scattering Times*

The elastic and transport scattering times for the surface carriers can be calculated from Eqs. (4.124) and (4.127), respectively. Since $\mathbf{e}_n$ is an odd function of the unit wave vector $\boldsymbol{n}$, there is no backscattering. The elastic life-time is independent of the warping and coincides with the disorder-related time scale, τ_0,

$$\tau = \tau_0. \tag{4.184}$$

However, the warping enters the transport scattering time through the angle average of the two anisotropic factors:

$$\frac{1}{\tau_{tr}} = \frac{1 - \overline{(\boldsymbol{n} \cdot \boldsymbol{n}_1)(\mathbf{e}_n \cdot \mathbf{e}_{n_1})}}{\tau_0} = \frac{1 - e_{\parallel}^2/2}{\tau_0} = \frac{1}{2\tau_0}\left[1 + \frac{\overline{\mathcal{M}_{k_F\cdot\boldsymbol{n}}^2}}{\mathcal{A}^2k_F^2}\right]. \tag{4.185}$$

This means that the conductivity corrections $\delta\sigma_1$ (4.174) and $\delta\sigma_2$ (4.175) also depend on the warping through the prefactor $(\tau_{tr}/\tau)^2$. We shall see, however, that in the net conductivity $\delta\sigma = \delta\sigma_1 + 2\delta\sigma_2$ the different warping terms cancel each other.

4.5.2 *Cooperon*

Our next step is to specify the Cooperon equations (4.145) and (4.146) for the surface state. Introducing the notations $\mathbf{e}_{\pm\boldsymbol{n}} = \mathbf{e}_\pm = \pm\mathbf{e}$ and using the identities

$$\mathbf{e}_- = -\mathbf{e}_+, \quad \mathbf{e}_+ - \mathbf{e}_- = 2\mathbf{e}, \quad \mathbf{e}_+ \cdot \mathbf{e}_- = -1, \quad \mathbf{e}_+ \times \mathbf{e}_- = 0, \tag{4.186}$$

we rewrite Eqs. (4.145) and (4.146) as

$$C^{0j} = \tau_0\delta_{0j} + \langle 1\rangle C^{0j} + \sum_{b=x,y,z} \langle \mathbf{e}\cdot\mathbf{b}\rangle C^{bj}, \tag{4.187}$$

$$C^{aj} = \tau_0\delta_{aj} + \langle \mathbf{e}\cdot\mathbf{a}\rangle C^{0j} + \sum_{b=x,y,z} \langle (\mathbf{e}\cdot\mathbf{a})(\mathbf{e}\cdot\mathbf{b})\rangle C^{bj}, \tag{4.188}$$

or, explicitly,

$$[1 - \langle 1\rangle]\, C^{0j} - \langle e_x\rangle C^{xj} - \langle e_y\rangle C^{yj} = \tau_0\delta_{0j}, \tag{4.189}$$

$$-\langle e_x\rangle C^{0j} + \left[1 - \langle e_x^2\rangle\right] C^{xj} = \tau_0\delta_{xj}, \tag{4.190}$$

$$-\langle e_y\rangle C^{0j} + \left[1 - \langle e_y^2\rangle\right] C^{yj} = \tau_0\delta_{yj}, \tag{4.191}$$

$$\left[1 - \langle e_\perp^2\rangle\right] C^{zj} = \tau_0\delta_{zj}. \tag{4.192}$$

Again, the coefficients in these equations can be evaluated by expanding the denominator in Eq. (4.138) and averaging over the directions of the unit wave vector $\boldsymbol{n}$. Then, for the diagonal Cooperons we find

$$C^{00}(\boldsymbol{q},\omega) = \frac{1}{D\boldsymbol{q}^2 - i\omega}, \qquad D = \frac{v_F^2\tau_{tr}}{2}, \tag{4.193}$$

$$\overline{C^{xx}} = \overline{C^{yy}} = \tau_{tr}\left(1 - \frac{e_\parallel^2}{4}\right) - \frac{e_\parallel^2}{4} i\tau_{tr}\omega C^{00}(\boldsymbol{q},\omega), \tag{4.194}$$

$$C^{zz} = \frac{\tau_0}{1 - \langle e_\perp^2\rangle}. \tag{4.195}$$

It is instructive to compare these results to those obtained for the TI thin film [see Eqs. (4.157)–(4.159)]. The most essential difference is the absence of the relaxation gap in the singlet Cooperon C^{00}, Eq. (4.193). This reflects the absence of the backscattering for the gapless TI surface states, irrespective of the warping. Although the latter has been treated perturbatively, it is possible to prove the gapless character of the singlet Cooperon for the warping of arbitrary strength [160], which is the consequence of the TRS. The warping only modifies the diffusion constant D through the transport scattering time (4.185) (see also Ref. [160]). We also note that the relevant contribution of the triplet Cooperons C^{xx} and C^{yy} [see Eq. (4.194)] is proportional to the small parameter $-i\omega\tau_{tr} \sim \tau_{tr}/\tau_\phi \ll 1$ and, therefore, can be neglected. The same is true for the other triplet Cooperon C^{zz}.

4.5.3 *Quantum Conductivity Correction*

We now turn to the conductivity correction. In order to calculate it, we need to do the angle averaging in Eqs. (4.174) and (4.175) for the surface-state spin helicity defined by the vector $\mathbf{e}_n$ in Eq. (4.183). Repeating the calculations outlined in subsection 4.4.9, we express the conductivity corrections $\delta\sigma_1$ and $\delta\sigma_2$ in terms of the diagonal Cooperons:

$$\delta\sigma_1 = \frac{e^2 D_0}{\pi\hbar}\left(\frac{\tau_{tr}}{\tau}\right)^2 \int \frac{d\boldsymbol{q}}{(2\pi)^2}\big[2\langle n_x^2\rangle C^{00} - 2\langle n_x^2 e_x^2\rangle C^{xx} - 2\langle n_x^2 e_y^2\rangle C^{yy} - 2\langle n_x^2 e_\perp^2\rangle C^{zz}\big], \tag{4.196}$$

$$\delta\sigma_2 = -\frac{e^2 D_0}{\pi\hbar}\left(\frac{\tau_{tr}}{\tau}\right)^2 \left(\frac{\mathbf{e}_\parallel}{2}\right)^2 \int \frac{d\boldsymbol{q}}{(2\pi)^2}\big[C^{00} - C^{xx}\big], \tag{4.197}$$

where $D_0 = v^2\tau/2$ involves the elastic life-time rather than τ_{tr} [*cf.* Eq. (4.193) for D]. Then, keeping only the singlet Cooperon C^{00}, we obtain the following expression for the net correction:

$$\delta\sigma_{xx} = \delta\sigma_{xx}^{(1)} + 2\delta\sigma_{xx}^{(2)} = \frac{e^2 D_0}{\pi\hbar}\left(\frac{\tau_{tr}}{\tau}\right)^2\left(1 - \frac{e_\parallel^2}{2}\right)\int \frac{d\boldsymbol{q}}{(2\pi)^2} C^{00}(\boldsymbol{q})$$
$$= \frac{e^2 D}{\pi\hbar}\int \frac{d\boldsymbol{q}}{(2\pi)^2} C^{00}(\boldsymbol{q}) = \frac{e^2}{2\pi h}\ln\frac{\tau_\phi}{\tau}. \tag{4.198}$$

The important observation here is that the warping-dependent parameters cancel each other [see also Eq. (4.185)]:

$$D_0\left(\frac{\tau_{tr}}{\tau}\right)^2\left(1-\frac{e_{\parallel}^2}{2}\right)=D\frac{\tau_{tr}}{\tau}\frac{\tau}{\tau_{tr}}=D, \tag{4.199}$$

so that $\delta\sigma$ retains the universal form typical of the symplectic symmetry class of disordered systems. This conclusion is independent of the warping strength [160].

4.5.4 *Magnetoconductivity*

So far we have treated the surface states in 3D TIs as purely two-dimensional. There is, however, a finite length λ of order of a few nanometer over which they penetrate into the bulk of the material. Since λ is quite small, the surface WAL conductivity is sensitive to the orientation of an external magnetic field with respect to the surface of the material. As demonstrated in Refs. [56, 148], such anisotropic B-field dependence can be used to detect the surface states in 3D TIs. Defining the magnetoconductivity as $\Delta\sigma(B)=\delta\sigma(B)-\delta\sigma(0)$, one can obtain the following B-field dependences for perpendicular ($\perp$) and parallel ($\parallel$) field orientations:

$$\Delta\sigma_{\perp}(B)=\frac{e^2}{2\pi h}\left[\ln\frac{B_{\perp}}{B}-\psi\left(\frac{1}{2}+\frac{B_{\perp}}{B}\right)\right],\quad B_{\perp}=\frac{\hbar}{4|e|\,\ell_{\varphi}^2}, \tag{4.200}$$

$$\Delta\sigma_{\parallel}(B)=-\frac{e^2}{2\pi h}\ln\left(1+\frac{B^2}{B_{\parallel}^2}\right),\quad B_{\parallel}=\frac{\hbar}{\sqrt{2}|e|\lambda\ell_{\varphi}^2}, \tag{4.201}$$

where $\psi(x)$ is the digamma function. The equations above contain the same phase-coherence length $\ell_{\varphi}=\sqrt{D\tau_{\varphi}}$. However, the magnetic-field scales, $B_{\perp}$ and $B_{\parallel}$, on which the magnetoconductivity decreases, are distinctly different. The field $B_{\perp}$ corresponds to the Aharonov–Bohm magnetic flux of order of the quantum h/e through a typical area $\sim\ell_{\varphi}^2$ swept by the interfering trajectories [123], while $B_{\parallel}$ corresponds to the same flux h/e, but through a significantly smaller area $\sim\lambda\ell_{\varphi}$ which is proportional to the surface-state penetration length into the bulk [148].

In order to identify the surface state one should extract its penetration length λ from Eqs. (4.200) and (4.201). Excluding ℓ_{φ}

from Eqs. (4.200) and (4.201), we can express λ in terms of two parameters, $B_\perp$ and $B_\parallel$, which can be obtained independently from fitting the corresponding experimental data [148]:

$$\lambda = \sqrt{\frac{2\hbar}{|e|}\frac{B_\perp}{B_\parallel^2}}. \tag{4.202}$$

The knowledge of the penetration length is of significant practical importance. It helps to estimate the critical thickness of the sample at which the two surface states start to overlap and the gapless TI state disappears.

To conclude this chapter, we should mention that the WAL in 3DTIs is suppressed if an energy gap opens in the surface spectrum as a result of TRS breaking [147]. This should apply to 3DTIs with magnetically doped surfaces (Ref. [146]). The suppressed WAL has also been found for ultrathin TIs films where it can be attributed to the overlap of the states from the opposite surfaces [150]. Besides, quantum transport in electrically gated TI films may depend sensitively on the coupling between surface and bulk states, which can lead to both WAL and WL regimes [161].

4.6 Problems

(1) Calculate the commutator $z\hat{\mathcal{H}} - \hat{\mathcal{H}}z$ for the Kane Hamiltonian in Eq. (4.2) and prove Eq. (4.5) for the normal component of the carrier velocity in HgTe.

(2) Consider a film of strained HgTe occupying the region $-d \leq z \leq 0$. The film thickness, d, is assumed much larger than the penetration depth λ of the surface state [see Eq. (4.26)],

$$d \gg \lambda. \tag{4.203}$$

Under this condition we can neglect the overlap of the surface states originating from $z = 0$ and $z = -d$ and treat them independently of each other. By analogy with the solution obtained in Section 4.1.2 show that the ansatz

$$\psi(x, y, z) = \Phi_{\boldsymbol{k}} e^{i(k_x x + k_y y) - (z+d)/\lambda}, \qquad \Phi_{\boldsymbol{k}} = -\sigma_0 \tau_y \Phi_{\boldsymbol{k}} \tag{4.204}$$

describes the surface state localized at $z = -d$. Find its spectrum, decay length and helicity. Note that this solution satisfies the boundary condition $\psi(x, y, -d) = -\sigma_0 \tau_y \psi(x, y, -d)$ where the sign is opposite to that in Eq. (4.19) for $z = 0$. Prove that the "top" and "bottom" surface states are orthogonal to each other.

(3) Consider the Hamiltonian for a TI thin film with the Fermi level lying in the conduction band [see Eqs. (4.40), (4.41) and Fig. 4.4]. Using the results of Secs. 2.1.1 and 2.1.2, calculate the Berry phases for the upper and lower blocks $\hat{H}_{\mathbf{k}\pm}$ at the Fermi level. Show that the elastic life-time τ in Eq. (4.126) can be expressed through the calculated Berry phases $\gamma_\pm$ as

$$\tau = \frac{\tau_0}{1 + (\gamma_+ - \pi)^2/\pi^2} = \frac{\tau_0}{1 + (\gamma_- - \pi)^2/\pi^2}. \tag{4.205}$$

In the same way, prove the following expression for the backscattering rate $\tau_{\mathcal{M}}^{-1}$ in the Cooperons and WAL conductivity [see Eqs. (4.157) and (4.180)]:

$$\tau_{\mathcal{M}}^{-1} = \frac{2}{\tau}\left(\frac{\gamma_+ - \pi}{\pi}\right)^2 = \frac{2}{\tau}\left(\frac{\gamma_- - \pi}{\pi}\right)^2. \tag{4.206}$$

Equations (4.205) and (4.206) demonstrate that the carrier backscattering occurs when the band Berry phase deviates from the universal value of π.

(4) Consider a 3DTI with an infinitely extended surface in the x-direction and a finite length L in the y-direction with the periodic boundary conditions at the ends. In this case, Eq. (4.198) for the WAL correction to the classical surface conductivity can be written as

$$\delta\sigma_{xx} = \frac{e^2 D}{\pi \hbar} \int \frac{dq_x}{2\pi L} \sum_n C^{00}(q_x, q_y^{(n)}), \tag{4.207}$$

where the Cooperon $C^{00}(q_x, q_y^{(n)})$ is given by Eq. (4.193) with $q_y^{(n)}$ taking the discrete values $2\pi n/L$ with integer $n = 0, \pm 1, \pm 2, \ldots$. Calculate the integral over q_x and show that under the condition

$$\ell_\varphi^2 \gg \left(\frac{L}{2\pi}\right)^2 \tag{4.208}$$

the WAL is effectively quasi-one-dimensional. Prove that for such a 3DTI "wire" the WAL correction (4.207) is

$$\delta\sigma_{xx} \approx \frac{e^2}{h} \frac{\ell_\varphi}{L}. \tag{4.209}$$

Chapter 5

Unconventional Superconductivity and Majorana Fermions in Topological Insulators

So far we have been dealing with the single-particle properties of the helical states in TIs. However, the spin helicity can play an important role in many-body phenomena as well. This chapter focuses on the interplay of the spin helicity and superconducting pairing in 3DTIs. This topic has attracted a great deal of interest for several reasons. First, superconductivity in TIs is unconventional. Unlike typical superconductors, for example Pb, Al, or Nb, where Cooper pairs have the ordinary s-wave symmetry, TIs can exhibit spin-triplet p-wave pair correlations. Second, superconducting TI structures support unconventional zero-energy bound states protected by the TRS. These bound states are the condensed matter analogues of Majorana fermions of relativistic field theory; hypothetical particles that are identical to their own antiparticles. Finally, the superconductivity in TIs can be induced through the proximity effect, by depositing conventional superconducting metals, such as Al or Nb, on the surface of a TI material. This offers divers possibilities for engineering topological superconducting states and devices.

Topological Insulators: The Physics of Spin Helicity in Quantum Transport
Grigory Tkachov

ISBN 978-981-4613-25-5 (Hardcover), 978-981-4613-26-2 (eBook)
www.panstanford.com

5.1 Theory of Superconductivity: A Primer

5.1.1 *Mean-Field Approximation. Superconducting Order Parameter*

The theory of superconductivity created by Bardeen, Cooper, and Schrieffer [206] attributes this phenomenon to phonon-mediated attractive interactions between pairs of electrons in a metal. An insightful phenomenological model of the superconducting pairing is an electron system with local attractive pair interactions described by the Hamiltonian [207]

$$H = \int d\boldsymbol{r} \sum_{\alpha,\beta=\uparrow,\downarrow} \left[\psi^{\dagger}_{\alpha}(\boldsymbol{r}) h_{\alpha\beta}(\boldsymbol{r})\psi_{\beta}(\boldsymbol{r}) + \frac{g}{2}\psi^{\dagger}_{\alpha}(\boldsymbol{r})\psi^{\dagger}_{\beta}(\boldsymbol{r})\psi_{\beta}(\boldsymbol{r})\psi_{\alpha}(\boldsymbol{r})\right], \tag{5.1}$$

where operator $\psi^{\dagger}_{\alpha}(\boldsymbol{r})$ ($\psi_{\alpha}(\boldsymbol{r})$) creates (annihilates) an electron at position $\boldsymbol{r}$ in spin state $\alpha = \uparrow, \downarrow$, $h_{\alpha\beta}(\boldsymbol{r})$ is the first-quantized form of the single-particle Hamiltonian, and $g < 0$ is the interaction constant. The electron motion is governed by the equation $i\hbar\partial_t\psi_{\alpha}(\boldsymbol{r}, t) = \psi_{\alpha}(\boldsymbol{r}, t)H - H\psi_{\alpha}(\boldsymbol{r}, t)$ or, explicitly,

$$i\hbar\partial_t\psi_{\alpha}(\boldsymbol{r}, t) = \sum_{\beta}\left[h_{\alpha\beta}(\boldsymbol{r})\psi_{\beta}(\boldsymbol{r}, t) - g\psi^{\dagger}_{\beta}(\boldsymbol{r}, t)\psi_{\alpha}(\boldsymbol{r}, t)\psi_{\beta}(\boldsymbol{r}, t)\right]. \tag{5.2}$$

In order to solve it we employ the mean-field approximation for the interaction term:

$$g\psi^{\dagger}_{\beta}(\boldsymbol{r}, t)\psi_{\alpha}(\boldsymbol{r}, t)\psi_{\beta}(\boldsymbol{r}, t) \to g\psi^{\dagger}_{\beta}(\boldsymbol{r}, t)\langle\psi_{\alpha}(\boldsymbol{r}, t)\psi_{\beta}(\boldsymbol{r}, t)\rangle \tag{5.3}$$

$$= -\Delta_{\alpha\beta}(\boldsymbol{r})\psi^{\dagger}_{\beta}(\boldsymbol{r}, t), \tag{5.4}$$

where the pair of the annihilation operators is replaced by its expectation value denoted by brackets $\langle\ldots\rangle$, and $\Delta_{\alpha\beta}(\boldsymbol{r})$ is the resulting mean-field potential given by

$$\Delta_{\alpha\beta}(\boldsymbol{r}) = -g\langle\psi_{\alpha}(\boldsymbol{r}, t)\psi_{\beta}(\boldsymbol{r}, t)\rangle, \qquad \alpha, \beta = \uparrow, \downarrow. \tag{5.5}$$

The expectation value $\langle\psi_{\alpha}(\boldsymbol{r}, t)\psi_{\beta}(\boldsymbol{r}, t)\rangle$ is called the superconducting order parameter. For an equilibrium system at zero temperature, the averaging is performed in the ground state. The nonzero order parameter indicates the specific character of the superconducting ground state which is formed by electron pairs (Cooper pairs) rather than by independent particles. For this reason,

$\Delta_{\alpha\beta}(\boldsymbol{r})$ is frequently called the pair potential. It has four components that can be combined into a 2×2 matrix

$$\hat{\Delta}(\boldsymbol{r}) = \begin{bmatrix} \Delta_{\uparrow\uparrow}(\boldsymbol{r}) & \Delta_{\uparrow\downarrow}(\boldsymbol{r}) \\ \Delta_{\downarrow\uparrow}(\boldsymbol{r}) & \Delta_{\downarrow\downarrow}(\boldsymbol{r}) \end{bmatrix} = \Delta(\boldsymbol{r}) \begin{bmatrix} 0 & 1 \\ -1 & 0 \end{bmatrix} = \Delta(\boldsymbol{r}) i\sigma_y, \qquad (5.6)$$

where we used the anticommutation relation $\psi_\alpha(\boldsymbol{r}, t)\psi_\beta(\boldsymbol{r}, t) = -\psi_\alpha(\boldsymbol{r}, t)\psi_\beta(\boldsymbol{r}, t)$, defining $\Delta(\boldsymbol{r})$ as

$$\Delta(\boldsymbol{r}) = -g\langle\psi_\uparrow(\boldsymbol{r}, t)\psi_\downarrow(\boldsymbol{r}, t)\rangle. \qquad (5.7)$$

Equation (5.6) corresponds to the conventional spin-singlet s-wave pairing. Indeed, the spin function of the electron pair, $i\sigma_y$, is antisymmetric under the exchange of the spin projections $\uparrow$ and $\downarrow$. Consequently, due to the Pauli principle, the orbital part of the pair potential, $\Delta(\boldsymbol{r})$, must be even under the exchange of the particle coordinates.

From Eqs. (5.2)–(5.4) we obtain the mean-field equation of motion

$$i\hbar\partial_t\psi_\alpha(\boldsymbol{r}, t) = \sum_\beta \left[h_{\alpha\beta}(\boldsymbol{r})\psi_\beta(\boldsymbol{r}, t) + \Delta_{\alpha\beta}(\boldsymbol{r})\psi_\beta^\dagger(\boldsymbol{r}, t)\right]. \qquad (5.8)$$

Since ψ and $\psi^\dagger$ are coupled, in order to solve Eq. (5.8) we must also use its hermitian conjugate:

$$i\hbar\partial_t\psi_\alpha^\dagger(\boldsymbol{r}, t) = \sum_\beta \left[-h_{\alpha\beta}^*(\boldsymbol{r})\psi_\beta^\dagger(\boldsymbol{r}, t) - \Delta_{\alpha\beta}^*(\boldsymbol{r})\psi_\beta(\boldsymbol{r}, t)\right]. \qquad (5.9)$$

At this stage it is convenient to introduce the spinor notations,

$$\psi(\boldsymbol{r}, t) = \begin{bmatrix} \psi_\uparrow(\boldsymbol{r}, t) \\ \psi_\downarrow(\boldsymbol{r}, t) \end{bmatrix}, \qquad \overline{\psi}(\boldsymbol{r}, t) = \begin{bmatrix} \psi_\uparrow^\dagger(\boldsymbol{r}, t) \\ \psi_\downarrow^\dagger(\boldsymbol{r}, t) \end{bmatrix}, \qquad (5.10)$$

and write Eqs. (5.8) and (5.9) as

$$i\hbar\partial_t \begin{bmatrix} \psi(\boldsymbol{r}, t) \\ \overline{\psi}(\boldsymbol{r}, t) \end{bmatrix} = \begin{bmatrix} \hat{h}(\boldsymbol{r}) & \hat{\Delta}(\boldsymbol{r}) \\ -\hat{\Delta}^*(\boldsymbol{r}) & -\hat{h}^*(\boldsymbol{r}) \end{bmatrix} \begin{bmatrix} \psi(\boldsymbol{r}, t) \\ \overline{\psi}(\boldsymbol{r}, t) \end{bmatrix}, \qquad (5.11)$$

where the hat over $h(\boldsymbol{r})$ indicates that the single-particle Hamiltonian is a 2×2 matrix in spin space. Thus, for the four-component spinor,

$$\Psi(\boldsymbol{r}, t) = \begin{bmatrix} \psi(\boldsymbol{r}, t) \\ \overline{\psi}(\boldsymbol{r}, t) \end{bmatrix} = \begin{bmatrix} \psi_\uparrow(\boldsymbol{r}, t) \\ \psi_\downarrow(\boldsymbol{r}, t) \\ \psi_\uparrow^\dagger(\boldsymbol{r}, t) \\ \psi_\downarrow^\dagger(\boldsymbol{r}, t) \end{bmatrix}, \qquad (5.12)$$

and the 4×4 matrix Hamiltonian,

$$\hat{H} = \begin{bmatrix} \hat{h}(\boldsymbol{r}) & \hat{\Delta}(\boldsymbol{r}) \\ -\hat{\Delta}^*(\boldsymbol{r}) & -\hat{h}^*(\boldsymbol{r}) \end{bmatrix}, \tag{5.13}$$

Eq. (5.11) has the form of a free-particle equation of motion:

$$i\hbar\partial_t \Psi(\boldsymbol{r}, t) = \hat{H}\,\Psi(\boldsymbol{r}, t). \tag{5.14}$$

In order to obtain this important result we have used the particle-hole (also called Nambu) representation for the field $\Psi(\boldsymbol{r}, t)$. The name stems from the fact that the spinors $\psi(\boldsymbol{r}, t)$ and $\overline{\psi}(\boldsymbol{r}, t)$ correspond to a quasi-particle and a quasi-hole fermionic excitations, respectively. In the normal (non-superconducting) case the quasi-particle is created by adding an electron into an empty state above the Fermi sea, whereas the quasi-hole is created by removing an electron from an occupied state of the Fermi sea. Since in normal metals these processes can occur independently, the particle and hole are described by the decoupled equations of motion,

$$\underbrace{i\hbar\partial_t\psi(\boldsymbol{r}, t) = \hat{h}(\boldsymbol{r})\psi(\boldsymbol{r}, t)}_{particle}, \quad \underbrace{i\hbar\partial_t\overline{\psi}(\boldsymbol{r}, t) = -\hat{h}^*(\boldsymbol{r})\overline{\psi}(\boldsymbol{r}, t)}_{hole}. \tag{5.15}$$

In a superconductor, however, an excited state appears to be a mixture of the particle and hole excitations of the underlying electron system. Still, the similarity of the mathematical structures of Eqs. (5.14) and (5.15) allows us to treat the superconducting and normal systems on equal footing, using the same methods. It is worth to mention another frequently used Nambu basis which is obtained from Eq. (5.12) by a unitary transformation,

$$\Psi'(\boldsymbol{r}, t) = \hat{U}\,\Psi(\boldsymbol{r}, t), \quad \hat{H}' = \hat{U}\,\hat{H}\,\hat{U}^\dagger, \quad \hat{U} = \begin{bmatrix} \sigma_0 & 0 \\ 0 & i\sigma_y \end{bmatrix}, \tag{5.16}$$

or, explicitly

$$\Psi'(\boldsymbol{r}, t) = \begin{bmatrix} \psi_\uparrow(\boldsymbol{r}, t) \\ \psi_\downarrow(\boldsymbol{r}, t) \\ \psi_\downarrow^\dagger(\boldsymbol{r}, t) \\ -\psi_\uparrow^\dagger(\boldsymbol{r}, t) \end{bmatrix}, \quad \hat{H}' = \begin{bmatrix} \hat{h}(\boldsymbol{r}) & \Delta(\boldsymbol{r})\sigma_0 \\ \Delta^*(\boldsymbol{r})\sigma_0 & -\mathbb{T}\hat{h}(\boldsymbol{r})\mathbb{T}^{-1} \end{bmatrix}. \tag{5.17}$$

In $\hat{H}'$, the hole Hamiltonian is related to the particle one by the TRS transformation with $\mathbb{T} = i\sigma_y C$, and the pair potential is diagonal is spin space.

5.1.2 Bogoliubov–de Gennes Equation. Particle-Hole Symmetry

In studies of superconducting systems (especially inhomogeneous ones), it is often convenient to deal with the wave functions instead of the field operators. In order to derive the equations for the wave functions we express the real-space field operators in the basis of new fermionic operators, γ_n and $\gamma_n^\dagger$ [208],

$$\psi_\alpha(\boldsymbol{r}, t) = \sum_n [u_{\alpha n}(\boldsymbol{r}, t)\gamma_n + v_{\alpha n}^*(\boldsymbol{r}, t)\gamma_n^\dagger], \tag{5.18}$$

$$\psi_\alpha^\dagger(\boldsymbol{r}, t) = \sum_n [v_{\alpha n}(\boldsymbol{r}, t)\gamma_n + u_{\alpha n}^*(\boldsymbol{r}, t)\gamma_n^\dagger], \tag{5.19}$$

where the sum runs over the states with quantum numbers n that may include discrete and/or continuous indices. The position and time dependence is transferred from the operators to the wave functions $u_{\alpha n}(\boldsymbol{r}, t)$ and $v_{\alpha n}(\boldsymbol{r}, t)$. The corresponding transformation of the Nambu basis is

$$\begin{bmatrix} \psi(\boldsymbol{r}, t) \\ \overline{\psi}(\boldsymbol{r}, t) \end{bmatrix} = \sum_n \left(\begin{bmatrix} u_n(\boldsymbol{r}, t) \\ v_n(\boldsymbol{r}, t) \end{bmatrix} \gamma_n + \begin{bmatrix} v_n^*(\boldsymbol{r}, t) \\ u_n^*(\boldsymbol{r}, t) \end{bmatrix} \gamma_n^\dagger \right), \tag{5.20}$$

where we introduced the spinors

$$u_n(\boldsymbol{r}, t) = \begin{bmatrix} u_{\uparrow n}(\boldsymbol{r}) \\ u_{\downarrow n}(\boldsymbol{r}) \end{bmatrix} e^{-iE_n t/\hbar}, \qquad v_n(\boldsymbol{r}, t) = \begin{bmatrix} v_{\uparrow n}(\boldsymbol{r}) \\ v_{\downarrow n}(\boldsymbol{r}) \end{bmatrix} e^{-iE_n t/\hbar}, \tag{5.21}$$

describing a state with energy E_n. Inserting Eq. (5.20) into Eq. (5.11) and collecting the coefficients at γ_n and $\gamma_n^\dagger$, we arrive at two equations

$$E_n \begin{bmatrix} u_n(\boldsymbol{r}) \\ v_n(\boldsymbol{r}) \end{bmatrix} = \begin{bmatrix} \hat{h}(\boldsymbol{r}) & \hat{\Delta}(\boldsymbol{r}) \\ -\hat{\Delta}^*(\boldsymbol{r}) & -\hat{h}^*(\boldsymbol{r}) \end{bmatrix} \begin{bmatrix} u_n(\boldsymbol{r}) \\ v_n(\boldsymbol{r}) \end{bmatrix}, \tag{5.22}$$

and

$$-E_n \begin{bmatrix} v_n^*(\boldsymbol{r}) \\ u_n^*(\boldsymbol{r}) \end{bmatrix} = \begin{bmatrix} \hat{h}(\boldsymbol{r}) & \hat{\Delta}(\boldsymbol{r}) \\ -\hat{\Delta}^*(\boldsymbol{r}) & -\hat{h}^*(\boldsymbol{r}) \end{bmatrix} \begin{bmatrix} v_n^*(\boldsymbol{r}) \\ u_n^*(\boldsymbol{r}) \end{bmatrix}, \tag{5.23}$$

Equation (5.22) is a wave-function analogue of Eq. (5.11), called the Bogolubov -de Gennes (BdG) equation. It comes in a pair with Eq. (5.23). This fact reflects a symmetry between the positive- and negative-energy states in a superconductor, known as the particle-hole symmetry. If, for example, $E_n > 0$, then for each positive-energy solution of Eq. (5.22) there is a negative-energy solution of Eq. (5.23). Mathematically, the particle hole-symmetry can be described by a conjugation operation

$$\mathbb{C} = \begin{bmatrix} 0 & \sigma_0 \\ \sigma_0 & 0 \end{bmatrix} C, \qquad \mathbb{C}^{-1} = \mathbb{C}. \tag{5.24}$$

where C denotes complex conjugation, and σ_0 is the unit matrix in spin space. It is easy to see that Eqs. (5.22) and (5.23) are related to each other by the transformation

$$\begin{bmatrix} v_n^*(\boldsymbol{r}) \\ u_n^*(\boldsymbol{r}) \end{bmatrix} = \mathbb{C} \begin{bmatrix} u_n(\boldsymbol{r}) \\ v_n(\boldsymbol{r}) \end{bmatrix}, \qquad \mathbb{C}\hat{H}\mathbb{C}^{-1} = -\hat{H}. \tag{5.25}$$

As an example illustrating the basic property of the superconducting state, let us calculate the energy spectrum of a superconductor with a spin-independent quadratic Hamiltonian,

$$\hat{h}(\boldsymbol{r}) = \left(-\frac{\hbar^2\nabla^2}{2m} - \mu\right)\sigma_0, \tag{5.26}$$

and the position-independent pairing, $\hat{\Delta}(\boldsymbol{r}) = const$ (μ is the Fermi energy). The system is translational invariant, that is, the quasiparticle momentum $\boldsymbol{p}$ is a good quantum number, and we seek the solutions of Eq. (5.22) in the form of plane waves $e^{i\,\boldsymbol{pr}/\hbar}$:

$$E_{\boldsymbol{p}} u_{\boldsymbol{p}} = \left(\frac{\boldsymbol{p}^2}{2m} - \mu\right) u_{\boldsymbol{p}} + \Delta i\sigma_y v_{\boldsymbol{p}} \tag{5.27}$$

$$E_{\boldsymbol{p}} v_{\boldsymbol{p}} = -\Delta^* i\sigma_y u_{\boldsymbol{p}} - \left(\frac{\boldsymbol{p}^2}{2m} - \mu\right) v_{\boldsymbol{p}}. \tag{5.28}$$

Another important symmetry is the invariance under spin rotation around the y axis. It allows us to choose the solutions to be the eigenstates of σ_y:

$$\sigma_y u_{s\boldsymbol{p}} = s u_{s\boldsymbol{p}}, \qquad \sigma_y v_{s\boldsymbol{p}} = s v_{s\boldsymbol{p}}, \tag{5.29}$$

with the eigenvalues $s = \pm 1$. In this basis the BdG equations become diagonal in spin space. Equating the determinant of the system to zero, we find the energy levels

$$E_{s\boldsymbol{p}} = \pm\sqrt{\left(\frac{\boldsymbol{p}^2}{2m} - \mu\right)^2 + |\Delta|^2}. \tag{5.30}$$

Note that they are degenerate with respect to the spin projection s. Equation (5.30) shows that the minimum energy required for creating an elementary excitation in a superconductor is $|\Delta|$. That is, the superconducting pairing leads to an energy gap between excited states and the ground state. Also, in accord with the above-discussed particle-hole symmetry, for each positive-energy level there is a symmetric negative-energy state. As we see below, the Dirac surface states of 3D TIs undergo a similar transformation in the presence of the superconducting pairing.

To conclude this introductory section, we should note that for an intrinsic superconductor the amplitude of the pair potential, $\Delta(\boldsymbol{r})$, must be determined self-consistently [207]. First, we find the wave functions $u_n(\boldsymbol{r})$ and $v_n(\boldsymbol{r})$ for given $\Delta(\boldsymbol{r})$ from the BdG equation (5.22). Then, we insert Eqs. (5.18) and (5.19) for the field operators into Eq. (5.7) and perform averaging with the statistical operator of the system. After this Eq. (5.7) turns into an explicit equation for $\Delta(\boldsymbol{r})$ as a function of the position vector $\boldsymbol{r}$, interaction constant g and other microscopic parameters. We leave the details of these calculations aside since in the following we focus on the externally induced, rather than intrinsic (i.e., interaction-induced) superconductivity.

5.2 Induced Surface Superconductivity in TIs

5.2.1 *Classification of Superconducting Pairing. Mixed s- and p-Wave Superconductivity*

The formalism described above can be used to study superconducting properties of TIs. We focus on a 3DTI with non-interacting surface carriers. The superconductivity in this system can be induced by depositing a conventional singlet s-wave superconductor on the TI surface, as depicted Fig. 5.1. Such a hybrid system was first considered by Fu and Kane in Ref. [209]. In the presence of the superconductor the TI surface state is described by a mean-field equation very similar to Eq. (5.13):

$$i\hbar\partial_t \begin{bmatrix} c_{\boldsymbol{p}}(t) \\ \bar{c}_{-\boldsymbol{p}}(t) \end{bmatrix} = \underbrace{\begin{bmatrix} v\boldsymbol{\sigma}\cdot\boldsymbol{p}-\mu & i\sigma_y\tilde{\Delta} \\ -i\sigma_y\tilde{\Delta}^* & v\boldsymbol{\sigma}^*\cdot\boldsymbol{p}+\mu \end{bmatrix}}_{\hat{H}} \begin{bmatrix} c_{\boldsymbol{p}}(t) \\ \bar{c}_{-\boldsymbol{p}}(t) \end{bmatrix}, \tag{5.31}$$

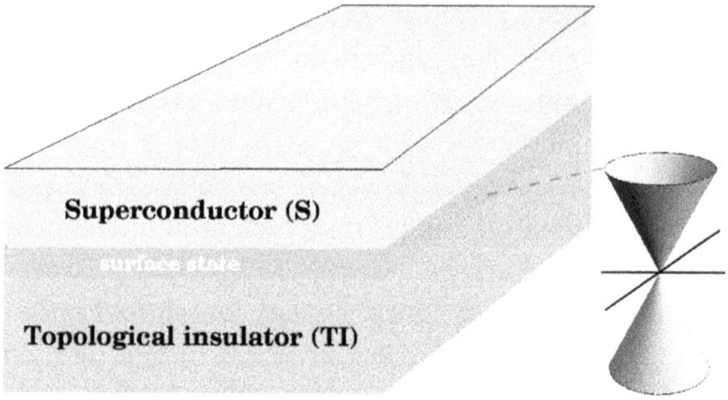

Figure 5.1 Schematic geometry of a superconductor (S)/topological insulator (TI) surface hybrid system. The surface state in the TI has a Dirac-cone band dispersion.

where $c_{\boldsymbol{p}}(t)$ and $\bar{c}_{-\boldsymbol{p}}(t)$ are the spinor field operators in the momentum ($\boldsymbol{p}$) representation

$$c_{\boldsymbol{p}}(t) = \begin{bmatrix} c_{\uparrow \boldsymbol{p}}(t) \\ c_{\downarrow \boldsymbol{p}}(t) \end{bmatrix}, \qquad \bar{c}_{-\boldsymbol{p}}(t) = \begin{bmatrix} c^{\dagger}_{\uparrow -\boldsymbol{p}}(t) \\ c^{\dagger}_{\downarrow -\boldsymbol{p}}(t) \end{bmatrix}, \tag{5.32}$$

and $v\boldsymbol{\sigma} \cdot \boldsymbol{p} - \mu$ is the Hamiltonian for the surface electrons (v is the Fermi velocity). The off-diagonal entries account for the induced singlet s-wave pairing with the strength $\tilde{\Delta}$. The induced $\tilde{\Delta}$ should not be confused with the pair potential of an intrinsic superconductor, Δ, introduced earlier in Eq. (5.7). The pair potential originates from the attractive electron–electron interaction and is proportional to the coupling constant $g < 0$. In the TI we assume no attractive interactions, that is, $g \geq 0$, which rules out the possibility of the intrinsic superconductivity. Hence, the pair potential is identically zero in the normal system. This, however, does not apply to the superconducting order parameter $\langle \psi_\alpha(\boldsymbol{r}, t)\psi_\beta(\boldsymbol{r}, t)\rangle$. It can appear in the normal system without any interactions due to the penetration of the Cooper pair condensate from the superconductor. This phenomenon is known as the superconducting proximity effect. The pairing $\tilde{\Delta}$ is related to the proximity-induced order parameter in the normal system. The proof of this fact requires a microscopic treatment of the coupling between the two systems. These aspects of the theory of the superconducting proximity effect in the TIs have been considered in Refs. [56, 119, 210–218].

As pointed out by Fu and Kane in Ref. [209], the usual singlet s-wave pairing in Eq. (5.31) induces unconventional triplet p-wave pair correlations. This happens because the spin-momentum locking $\boldsymbol{\sigma} \cdot \boldsymbol{p}$ violates the spin-rotation symmetry of the Hamiltonian in Eq. (5.31). In order to see this, and for a rigorous classification of the symmetry of the pair correlations, one has to define and calculate a suitable correlation function of the system. The standard way is to use the Green function of Eq. (5.31). It is defined as follows

$$\hat{G}_{\boldsymbol{p}}(t,t') = \frac{1}{i\hbar}\left\langle \begin{bmatrix} c_{\boldsymbol{p}}(t) \\ \bar{c}_{-\boldsymbol{p}}(t) \end{bmatrix} \otimes \left[\bar{c}^T_{\boldsymbol{p}}(t')\; c^T_{-\boldsymbol{p}}(t')\right] \right\rangle \tag{5.33}$$

$$= \begin{bmatrix} \langle c_{\boldsymbol{p}}(t) \otimes \bar{c}^T_{\boldsymbol{p}}(t')\rangle & \langle c_{\boldsymbol{p}}(t) \otimes c^T_{-\boldsymbol{p}}(t')\rangle \\ \langle \bar{c}_{-\boldsymbol{p}}(t) \otimes \bar{c}^T_{\boldsymbol{p}}(t')\rangle & \langle \bar{c}_{-\boldsymbol{p}}(t) \otimes c^T_{-\boldsymbol{p}}(t')\rangle \end{bmatrix}, \tag{5.34}$$

where symbol $\otimes$ means the direct products of the spinors in Eq. (5.32) and the transposed spinors

$$\bar{c}^T_{\boldsymbol{p}}(t') = [c^{\dagger}_{\uparrow\boldsymbol{p}}(t'), c^{\dagger}_{\downarrow\boldsymbol{p}}(t')], \qquad c^T_{-\boldsymbol{p}}(t') = [c_{\uparrow-\boldsymbol{p}}(t'), c_{\downarrow-\boldsymbol{p}}(t')], \tag{5.35}$$

and the brackets $\langle \ldots \rangle$ denote averaging of the time-ordered operator pairs with the ground-state statistical operator. Equation (5.34) defines the Green function as a 2×2 matrix in the particle-hole space. For the matrix elements in the particle-hole space we introduce the notations

$$\hat{G}_{\boldsymbol{p}}(t,t') = \begin{bmatrix} \hat{G}_{11,\boldsymbol{p}}(t,t') & \hat{G}_{12,\boldsymbol{p}}(t,t') \\ \hat{G}_{21,\boldsymbol{p}}(t,t') & \hat{G}_{22,\boldsymbol{p}}(t,t') \end{bmatrix}, \tag{5.36}$$

where each entry is a 2×2 matrix in the spin space. The pair correlations are described by the off-diagonal Green functions with equal time arguments, for example,

$$\hat{G}_{12,\boldsymbol{p}}(t,t) = \frac{1}{i\hbar}\begin{bmatrix} \langle c_{\uparrow\boldsymbol{p}}(t)c_{\uparrow-\boldsymbol{p}}(t)\rangle & \langle c_{\uparrow\boldsymbol{p}}(t)c_{\downarrow-\boldsymbol{p}}(t)\rangle \\ \langle c_{\downarrow\boldsymbol{p}}(t)c_{\uparrow-\boldsymbol{p}}(t)\rangle & \langle c_{\downarrow\boldsymbol{p}}(t)c_{\downarrow-\boldsymbol{p}}(t)\rangle \end{bmatrix}. \tag{5.37}$$

This equation encodes all pairing scenarios allowed by the fermionic anti-commutation rules. Since the total spin of an electron pair can be either 0 (singlet state) or 1 (triplet state), Eq. (5.37) can be expanded in the basis of the singlet and triplet spin states:

$$\hat{G}_{12,\boldsymbol{p}}(t,t) = \frac{1}{i\hbar}\left[d_0(\boldsymbol{p})i\sigma_y + \boldsymbol{d}(\boldsymbol{p}) \cdot \boldsymbol{\sigma} i\sigma_y\right]. \tag{5.38}$$

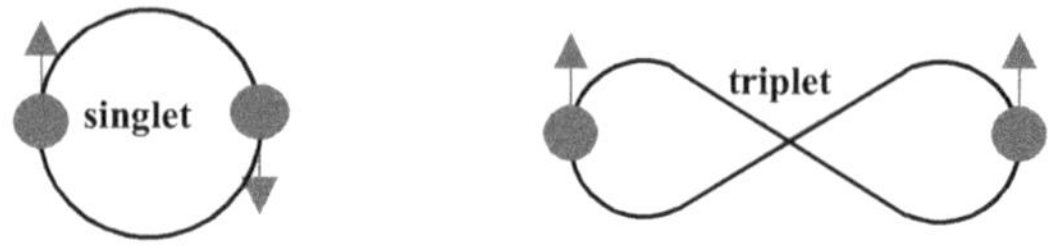

Figure 5.2 Illustration of the singlet and triplet pairing (see also the text).

The first term is the singlet part. It has the antisymmetric spin function, $i\sigma_y$, and the symmetric (under $\boldsymbol{p} \to -\boldsymbol{p}$) orbital function [*cf.* Eq. (5.6)]

$$d_0(\boldsymbol{p}) = \frac{1}{2}\langle c_{\uparrow \boldsymbol{p}}(t)c_{\downarrow -\boldsymbol{p}}(t) - c_{\downarrow \boldsymbol{p}}(t)c_{\uparrow -\boldsymbol{p}}(t)\rangle. \tag{5.39}$$

The second term in Eq. (5.38) is the triplet part. The triplet basis spin functions $\sigma_j i\sigma_y$ (where $j = x, y,$ or z) are symmetric under exchange of the spin indices. The corresponding orbital functions are combined into a vector $\boldsymbol{d}(\boldsymbol{p}) = [d_x(\boldsymbol{p}), d_y(\boldsymbol{p}), d_z(\boldsymbol{p})]$ defined by

$$d_x(\boldsymbol{p}) = -\frac{1}{2}\langle c_{\uparrow \boldsymbol{p}}(t)c_{\uparrow -\boldsymbol{p}}(t) - c_{\downarrow \boldsymbol{p}}(t)c_{\downarrow -\boldsymbol{p}}(t)\rangle, \tag{5.40}$$

$$d_y(\boldsymbol{p}) = -\frac{i}{2}\langle c_{\uparrow \boldsymbol{p}}(t)c_{\uparrow -\boldsymbol{p}}(t) + c_{\downarrow \boldsymbol{p}}(t)c_{\downarrow -\boldsymbol{p}}(t)\rangle, \tag{5.41}$$

$$d_z(\boldsymbol{p}) = \frac{1}{2}\langle c_{\uparrow \boldsymbol{p}}(t)c_{\downarrow -\boldsymbol{p}}(t) + c_{\downarrow \boldsymbol{p}}(t)c_{\uparrow -\boldsymbol{p}}(t)\rangle. \tag{5.42}$$

All three triplet orbital functions are odd in momentum $\boldsymbol{p}$. Unlike the singlet state, the triplet pairing is nodal in both momentum and coordinate spaces, for example, vanishing for two electrons at the same position, as illustrated in Fig. 5.2. Both the singlet and triplet pairings are consistent with the fermionic anti-commutation rules and the Pauli exclusion principle. It is worth noting that the states with $d_{x,y}(\boldsymbol{p})$ are formed by Cooper pairs with nonzero (± 1) projections of the total spin, whereas $d_z(\boldsymbol{p})$ corresponds to a state with zero total spin projection.

In order to identify the pairing symmetry in the TI we need to calculate $\hat{G}_{12,\boldsymbol{p}}(t, t)$ explicitly and compare it with the expansion in Eq. (5.38). We start with the equation of motion for the Green function (5.33)

$$i\hbar\partial_t \hat{G}_{\boldsymbol{p}}(t, t') - \hat{H}\hat{G}_{\boldsymbol{p}}(t, t') = \hat{I}\delta(t - t'). \tag{5.43}$$

It follows from Eqs. (5.31) and (5.33) by differentiating $\hat{G}_{\boldsymbol{p}}(t, t')$ with respect to the first time argument ($\hat{I}$ is the 4×4 unit matrix).

Using the energy representation

$$\hat{G}_{\boldsymbol{p}}(t,t') = \int \frac{dE}{h} \hat{G}_{\boldsymbol{p}E}\, e^{-iE(t-t')/\hbar}, \tag{5.44}$$

and Eqs. (5.31) and (5.36) for $\hat{H}$ and $\hat{G}_{\boldsymbol{p}E}$, respectively, we rewrite Eq. (5.43) as

$$\left\{ E\hat{I} - \begin{bmatrix} v\boldsymbol{\sigma}\cdot\boldsymbol{p} - \mu & i\sigma_y\tilde{\Delta} \\ -i\sigma_y\tilde{\Delta}^* & v\boldsymbol{\sigma}^*\cdot\boldsymbol{p} + \mu \end{bmatrix} \right\} \begin{bmatrix} \hat{G}_{11,\boldsymbol{p}E} & \hat{G}_{12,\boldsymbol{p}E} \\ \hat{G}_{21,\boldsymbol{p}E} & \hat{G}_{22,\boldsymbol{p}E} \end{bmatrix} = \hat{I}. \tag{5.45}$$

It splits into two pairs of coupled equations

$$(E + \mu - v\boldsymbol{\sigma}\boldsymbol{p})\hat{G}_{11,\boldsymbol{p}E} - \tilde{\Delta} i\sigma_y \hat{G}_{21,\boldsymbol{p}E} = \sigma_0, \tag{5.46}$$

$$\tilde{\Delta}^* i\sigma_y \hat{G}_{11,\boldsymbol{p}E} + (E - \mu - v\boldsymbol{\sigma}^*\boldsymbol{p})\hat{G}_{21,\boldsymbol{p}E} = 0, \tag{5.47}$$

and

$$(E + \mu - v\boldsymbol{\sigma}\boldsymbol{p})\hat{G}_{12,\boldsymbol{p}E} - \tilde{\Delta} i\sigma_y \hat{G}_{22,\boldsymbol{p}E} = 0, \tag{5.48}$$

$$\tilde{\Delta}^* i\sigma_y \hat{G}_{12,\boldsymbol{p}E} + (E - \mu - v\boldsymbol{\sigma}^*\boldsymbol{p})\hat{G}_{22,\boldsymbol{p}E} = \sigma_0. \tag{5.49}$$

To illustrate the solution procedure let us consider the first pair of the equations. Applying $\sigma_y(E - \mu - v\boldsymbol{\sigma}^*\boldsymbol{p})\sigma_y$ to Eq. (5.46) from the left and excluding $\hat{G}_{21,\boldsymbol{p}E}$ with the help of Eq. (5.47), we obtain a closed equation for $\hat{G}_{11,\boldsymbol{p}E}$

$$\left[(E - \mu + v\boldsymbol{\sigma}\boldsymbol{p})(E + \mu - v\boldsymbol{\sigma}\boldsymbol{p}) - |\tilde{\Delta}|^2\right] \hat{G}_{11,\boldsymbol{p}E} = E - \mu + v\boldsymbol{\sigma}\boldsymbol{p}. \tag{5.50}$$

Then, applying $(E + \mu + v\boldsymbol{\sigma}\boldsymbol{p})(E - \mu - v\boldsymbol{\sigma}\boldsymbol{p}) - |\tilde{\Delta}|^2$ from the left, with algebra the solution is

$$\hat{G}_{11,\boldsymbol{p}E} = \frac{(E + \mu + v\boldsymbol{\sigma}\boldsymbol{p})[(E - \mu)^2 - v^2p^2] - |\tilde{\Delta}|^2(E - \mu + v\boldsymbol{\sigma}\boldsymbol{p})}{\left[E^2 - (vp - \mu)^2 - |\tilde{\Delta}|^2\right]\left[E^2 - (vp + \mu)^2 - |\tilde{\Delta}|^2\right]}, \tag{5.51}$$

where $p \equiv |\boldsymbol{p}|$. The function $\hat{G}_{21,\boldsymbol{p}E}$ is then obtained from Eq. (5.47) as

$$\hat{G}_{21,\boldsymbol{p}E} = \frac{-\tilde{\Delta}^* i\sigma_y\,(E^2 - \mu^2 - v^2p^2 - |\tilde{\Delta}|^2 - 2\mu v\boldsymbol{\sigma}\boldsymbol{p})}{\left[E^2 - (vp - \mu)^2 - |\tilde{\Delta}|^2\right]\left[E^2 - (vp + \mu)^2 - |\tilde{\Delta}|^2\right]}. \tag{5.52}$$

Similarly, the solutions of Eqs. (5.48) and (5.49) are

$$\hat{G}_{22,\boldsymbol{p}E} = \frac{(E - \mu + v\boldsymbol{\sigma}^*\boldsymbol{p})[(E + \mu)^2 - v^2p^2] - |\tilde{\Delta}|^2(E + \mu + v\boldsymbol{\sigma}^*\boldsymbol{p})}{\left[E^2 - (vp - \mu)^2 - |\tilde{\Delta}|^2\right]\left[E^2 - (vp + \mu)^2 - |\tilde{\Delta}|^2\right]}, \tag{5.53}$$

$$\hat{G}_{12,\boldsymbol{p}E} = \frac{(\overbrace{E^2 - \mu^2 - v^2p^2 - |\tilde{\Delta}|^2}^{singlet} - \overbrace{2\mu v\boldsymbol{\sigma p}}^{triplet})\,\tilde{\Delta} i\sigma_y}{\left[E^2 - (vp-\mu)^2 - |\tilde{\Delta}|^2\right]\left[E^2 - (vp+\mu)^2 - |\tilde{\Delta}|^2\right]}. \tag{5.54}$$

The latter equation provides the information about the character of the pair correlations in the TI. Comparing Eqs. (5.38) and (5.54) we see that the induced superconductivity has both singlet ($\propto i\sigma_y$) and triplet ($\propto \sigma_j i\sigma_y$) parts with the corresponding orbital functions, $d_0(\boldsymbol{p}, E)$ and $\boldsymbol{d}(\boldsymbol{p}, E)$, given by

$$d_0(\boldsymbol{p}, E) = \frac{\tilde{\Delta}(E^2 - \mu^2 - v^2p^2 - |\tilde{\Delta}|^2)}{\left[E^2 - (vp-\mu)^2 - |\tilde{\Delta}|^2\right]\left[E^2 - (vp+\mu)^2 - |\tilde{\Delta}|^2\right]}, \tag{5.55}$$

$$\boldsymbol{d}(\boldsymbol{p}, E) = -\frac{2\tilde{\Delta}\mu v\boldsymbol{p}}{\left[E^2 - (vp-\mu)^2 - |\tilde{\Delta}|^2\right]\left[E^2 - (vp+\mu)^2 - |\tilde{\Delta}|^2\right]}. \tag{5.56}$$

Comparing these results with the case of the ordinary s-wave superconductors (see Section 5.1), we conclude that the induced superconductivity in the TI has an unconventional, mixed s- and p-wave character. The p-wave correlations occur as the combined effect of the induced s-wave pairing and the spin-momentum locking $\boldsymbol{\sigma} \cdot \boldsymbol{p}$, which breaks the spin-rotation symmetry. As expected, the orbital triplet pairing is odd in momentum $\boldsymbol{p}$. It has only two nonzero components, d_x and d_y. The third, d_z, is zero by the TRS. Besides, the triplet pairing vanishes for $\mu = 0$, that is, at the charge neutrality point of the surface state.

Apart from the pairing symmetry, Eqs. (5.55) and (5.56) yield the spectrum of quasiparticle excitations in the superconducting state. The excitation energies correspond to the zeros of the denominator. There are two spectral branches

$$E^{(1)}_{\boldsymbol{p}} = \pm\sqrt{(vp-\mu)^2 + |\tilde{\Delta}|^2}, \qquad E^{(2)}_{\boldsymbol{p}} = \pm\sqrt{(vp+\mu)^2 + |\tilde{\Delta}|^2}. \tag{5.57}$$

These are shown in Fig. 5.3. We see that, similar to the conventional superconductors [*cf.* Eq. (5.30)], there is an excitation gap $|\tilde{\Delta}|$ at the Fermi surface $p = p_F = \mu/v$.

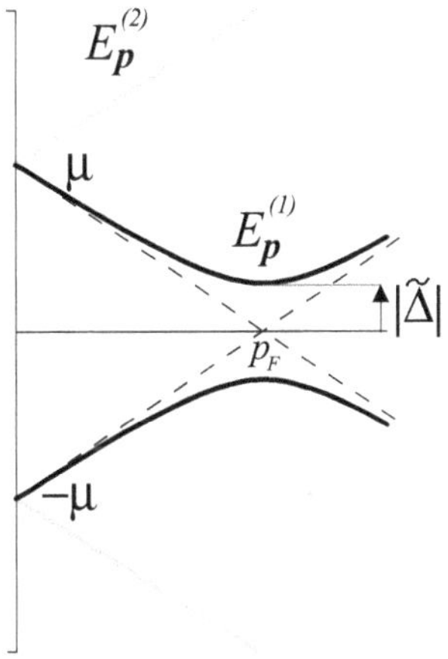

Figure 5.3 Energy spectrum of a superconducting TI from Eq. (5.57). The dashed lines indicate the normal-state particle ($vp - \mu$) and hole ($\mu - vp$) dispersions.

5.2.2 *Andreev Reflection. Topological Bound States in S/TI/S Junctions*

At this stage, one might ask: what are the potentially observable phenomena that could serve as evidence for the unconventional superconductivity in TIs? Here we address this question, following the paper of Fu and Kane, Ref. [209]. They proposed that the unconventional superconductivity in TIs should manifest itself in the formation of topological bound states at the boundaries between superconducting and normal TI phases [209]. A typical example of such a hybrid system is a Josephson junction created by depositing two spatially separated superconductors (Ss) on the surface of a 3DTI, as sketched in Fig. 5.4. The junctions of this type have been realized and investigated experimentally in Refs. [119, 120, 219–225]

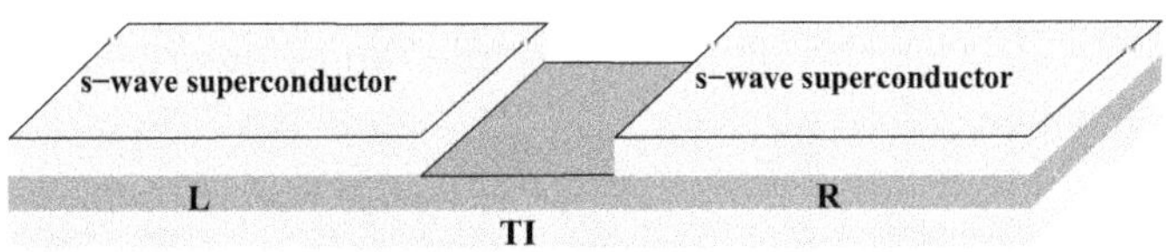

Figure 5.4 Schematic of a superconducting TI junction.

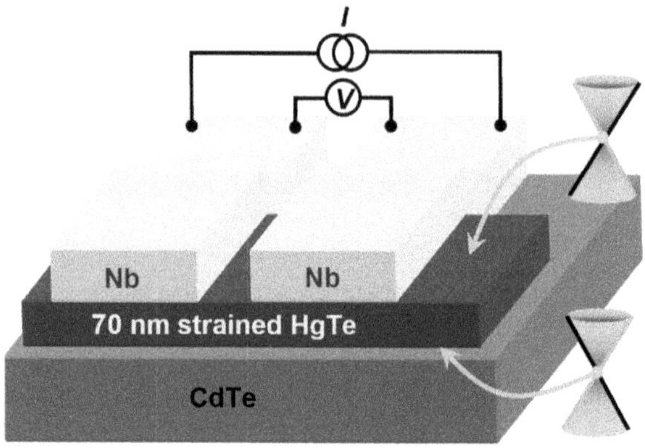

Figure 5.5 Schematic experimental realization of a superconducting junction with a 70-nm-thick strained HgTe film as a 3DTI and two Nb contacts (adapted from Ref. [120]).

(see also Fig. 5.5). We shall see that the S/TI/S junctions support protected states that, at the same time, are capable of carrying electric current without dissipation.

In the junction, the pairing $\tilde{\Delta}$ is nonuniform in space. It vanishes in the normal region and is nonzero in the left (L) and right (R) S terminals

$$\tilde{\Delta}|_{L,R} = \Delta_0 e^{i\phi_{L,R}}, \quad \phi_L = 0, \quad \phi_R \equiv \phi. \tag{5.58}$$

The pairing term is assumed to have different phases $\phi_{L,R}$ and the same absolute value $\Delta_0 = |\tilde{\Delta}|_{L,R}$ in terminals L and R. The junction length defined as distance between the S terminals is assumed much smaller than the superconducting coherence length $\sim \hbar v/\Delta_0$. For example, for Nb/TI junctions with typical parameters $\Delta_0 \approx 0.2$ meV and $\hbar v \approx 250 \div 350$ meV nm (Ref. [119]), the junction length should be smaller than $1.25 \div 1.75$ μm, which is easily realizable experimentally. In this limit, the junction can be described by the BdG equation

$$\left[E\hat{I} - \begin{pmatrix} -i\hbar v\boldsymbol{\sigma}\nabla + U(x) - \mu & i\sigma_y\Delta_0 e^{i\phi_{L,R}} \\ -i\sigma_y\Delta_0 e^{-i\phi_{L,R}} & -i\hbar v\boldsymbol{\sigma}^*\nabla - U(x) + \mu \end{pmatrix}\right]\begin{bmatrix} u(\boldsymbol{r}) \\ v(\boldsymbol{r}) \end{bmatrix}, \tag{5.59}$$

or, explicitly,

$$\begin{bmatrix} E+\mu-U(x) & \hbar v(i\partial_x+\partial_y) & 0 & -\Delta_0 e^{i\phi_{L,R}} \\ \hbar v(i\partial_x-\partial_y) & E+\mu-U(x) & \Delta_0 e^{i\phi_{L,R}} & 0 \\ 0 & \Delta_0 e^{-i\phi_{L,R}} & E-\mu+U(x) & \hbar v(i\partial_x-\partial_y) \\ -\Delta_0 e^{-i\phi_{L,R}} & 0 & \hbar v(i\partial_x+\partial_y) & E-\mu+U(x) \end{bmatrix} \times \begin{bmatrix} u_\uparrow(\mathbf{r}) \\ u_\downarrow(\mathbf{r}) \\ v_\uparrow(\mathbf{r}) \\ v_\downarrow(\mathbf{r}) \end{bmatrix} = 0. \quad (5.60)$$

To account for the possibility of electron scattering in the junction we introduce a potential $U(x) = U\delta(x)$ between the terminals. The precise form of the scattering potential is not of crucial importance here because, as we see below, the main interesting effects originate from Klein tunneling that takes place independently of the barrier shape [56, 226, 227].

We are interested in the particular solutions of the BdG equations—Andreev bound states (ABSs)—that have energies smaller than the induced excitation gap, Δ_0. The mechanism of the formation of the bound states, Andreev reflection, is illustrated in Fig. 5.6. Consider a particle with an energy $E < \Delta_0$ above the Fermi level propagating in the normal region toward the superconductor on the right. Since below the gap there are only the pair states, the particle entering the superconductor must be accompanied by

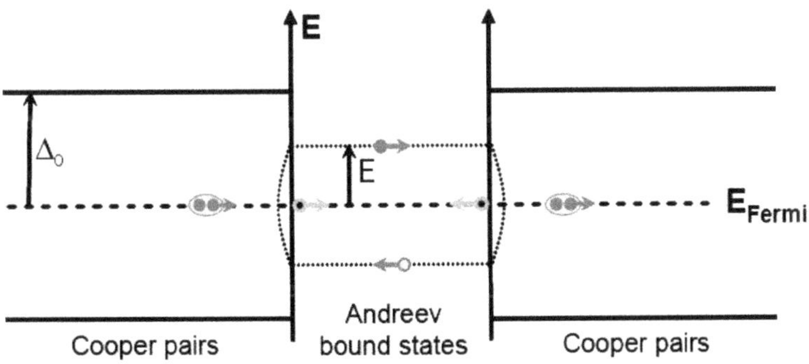

Figure 5.6 Illustration of Andreev reflection and formation of Andreev bound states in a superconductor/normal metal/superconductor junction (adapted from Ref. [120]).

another one with energy $-E$ to create a Cooper pair in the ground state at the Fermi level. This leaves a hole at energy $-E$ in the Fermi sea of the normal system, and by the TRS, the hole must retrace the trajectory of the particle. This is the Andreev reflection process. In order to enter the superconductor on the left the hole experiences another Andreev reflection, being emitted back into the normal region as a particle with energy E. In this way a bound state is formed between the two superconductors.

We seek the ABS solutions as the sum of independent modes with different k_y wave numbers:

$$\Phi(\boldsymbol{r}) \equiv \begin{bmatrix} u_\uparrow(\boldsymbol{r}) \\ u_\downarrow(\boldsymbol{r}) \\ v_\uparrow(\boldsymbol{r}) \\ v_\downarrow(\boldsymbol{r}) \end{bmatrix} = \sum_{k_y} \Phi_{k_y}(x) \frac{\mathrm{e}^{ik_y y}}{\sqrt{W}}, \tag{5.61}$$

where W is the junction width, and $\Phi_{k_y}(x)$ is the evanescent spinor function decaying for $x \to \pm\infty$. For a large Fermi energy

$$\mu \gg \Delta_0, \tag{5.62}$$

the spinors $\Phi_{k_y}(x)$ are readily obtained from Eq. (5.60) as

$$\Phi_{k_y}(x>0) = \begin{bmatrix} 1 \\ \mathrm{e}^{i\theta} \\ -\mathrm{e}^{i\theta} a_R^+ \\ a_R^+ \end{bmatrix} C_R^> \, \mathrm{e}^{ixk_F \cos\theta - x/\xi} + \begin{bmatrix} 1 \\ -\mathrm{e}^{-i\theta} \\ \mathrm{e}^{-i\theta} a_R^- \\ a_R^- \end{bmatrix} C_R^< \, \mathrm{e}^{-ixk_F \cos\theta - x/\xi}, \tag{5.63}$$

and

$$\Phi_{k_y}(x<0) = \begin{bmatrix} 1 \\ \mathrm{e}^{i\theta} \\ -\mathrm{e}^{i\theta} a_L^- \\ a_L^- \end{bmatrix} C_L^> \, \mathrm{e}^{ixk_F \cos\theta + x/\xi} + \begin{bmatrix} 1 \\ -\mathrm{e}^{-i\theta} \\ \mathrm{e}^{-i\theta} a_L^+ \\ a_L^+ \end{bmatrix} C_L^< \, \mathrm{e}^{-ixk_F \cos\theta + x/\xi},$$

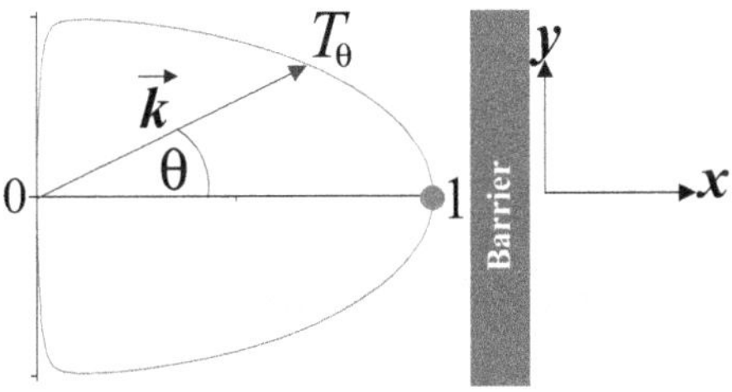

Figure 5.7 In-plane scattering geometry and polar plot of barrier transparency T_θ as a function of incidence angle θ [see also Eq. (5.70)].

where $C^{>,<}_{R,L}$ are constants, ξ is the length of the penetration of the ABSs into the superconductors:

$$\xi = \frac{\hbar v}{\sqrt{\Delta_0^2 - E^2}}, \tag{5.65}$$

and other parameters and notations are explained below:

$$a^{\pm}_{R,L}(E) = \frac{\Delta_0 e^{-i\phi_{R,L}}}{E \pm i\sqrt{\Delta_0^2 - E^2}}, \tag{5.66}$$

$$\cos\theta = \sqrt{1 - (k_y/k_F)^2}. \tag{5.67}$$

Here $a^{\pm}_{R,L}(E)$ are the amplitudes of Andreev reflection from superconductors R and L, the angle θ indicates the propagation direction, spanning the range from $-\pi/2$ to $\pi/2$ around the x-axis, as shown in Fig. 5.7. The allowed values of θ are determined by the condition that the transverse wave number k_y is smaller than the Fermi wave number k_F. The matching condition for spinors (5.63) and (5.64) is obtained by integrating Eq. (5.59) around $x = 0$, which yields

$$\Phi_{k_y}(-0) = \begin{bmatrix} 1 + iZ\sigma_x & 0 \\ 0 & 1 - iZ\sigma_x \end{bmatrix} \Phi_{k_y}(+0), \quad Z = \frac{U}{\hbar v}, \tag{5.68}$$

where the dimensionless parameter Z characterizes the barrier strength. Inserting (5.63) and (5.64) into (5.68) we arrive at four linear algebraic equations for the four constants $C^{>}_{R}$, $C^{<}_{R}$, $C^{>}_{L}$, and $C^{<}_{L}$. Equating the expressions the determinant of the system to zero, we arrive to the following equation for the ABSs:

$$[a^{+}_{R}(E) - a^{-}_{R}(E)][a^{+}_{L}(E) - a^{-}_{L}(E)] = T_\theta\,[a^{+}_{R}(E) - a^{+}_{L}(E)][a^{-}_{R}(E) - a^{-}_{L}(E)]. \tag{5.69}$$

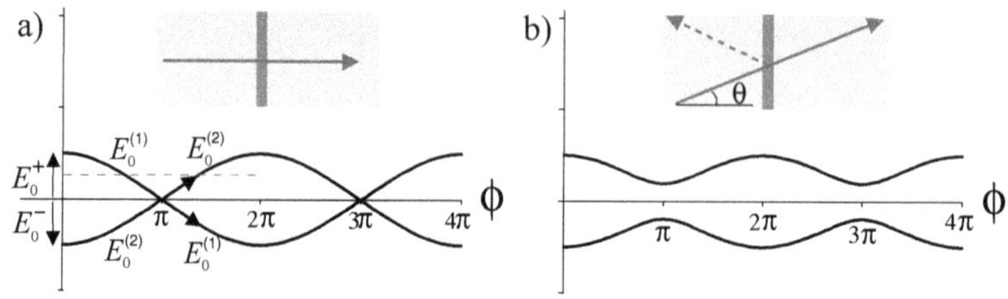

Figure 5.8 Andreev bound states (5.71) versus phase difference ϕ for (a) normal incidence at potential barrier with $\theta = 0$ and (b) oblique incidence with $\theta = \pi/4$. Arrows indicate the 4π-periodic branches (5.73). Dashed line corresponds to fixed energy (5.74).

Here the carrier helicity is accounted for by the angle-dependent transmission coefficient $T(\theta)$ (see also Fig. 5.7):

$$T_\theta = \frac{\cos^2(\theta)}{1 - \sin^2(\theta)/(1 + Z^2)}. \tag{5.70}$$

For the perpendicular propagation with $\theta = 0$, the transparency T_0 is 1 independently of the barrier strength. This is the familiar Klein tunneling of Dirac fermions. In the following we discuss how this phenomenon manifests itself in the energy spectrum of the ABSs.

Using Eq. (5.66) for the Andreev amplitudes, we obtain from (5.69) the ABS spectrum as a function of the superconducting phase difference ϕ and the propagation angle θ:

$$E_\theta^\pm(\phi) = \pm\Delta_0 \sqrt{1 - T_\theta \sin^2 \frac{\phi}{2}}. \tag{5.71}$$

Figure 5.8 shows the ABSs for two distinct cases: normal incidence at the potential barrier with $\theta = 0$ (panel a) and oblique incidence with $\theta \neq 0$ (panel b). As the perpendicular propagation is reflectionless, the ABSs for $\theta = 0$ are given by

$$E_0^\pm(\phi) = \pm\Delta_0 \left|\cos \frac{\phi}{2}\right|. \tag{5.72}$$

For the phase difference $\phi = \pi$ the two $\pm$ states have zero energy, that is, they are gapless. We recall that gapless ABSs also occur in ballistic junctions with conventional s-wave superconductors (Ref. [228]). However, unlike the latter, the gapless solutions (5.72) are protected by the TRS. In order to prove this we notice that the

$\pm$ states in Eq. (5.72) are equivalent to two 4π-periodic branches crossing in the middle of the gap:

$$E_0^{(1,2)}(\phi) = \pm\Delta_0 \cos\frac{\phi}{2}, \qquad (5.73)$$

where we introduced notation $E_0^{(1,2)}(\phi)$ to distinguish the 4π-periodic solutions (5.73) from (5.72). Let us now examine their wave functions at the fixed energy [shown by the dashed line in Fig. 5.8(a)], which implies

$$E_0^{(1)}(\phi) = E_0^{(2)}(2\pi - \phi) = E. \qquad (5.74)$$

The wave functions of these states [see, e.g., Eq. (5.63) for $\theta = 0$] appear to consist only of one-way propagating states:

$$\Phi_0^{(1)}(x) = \begin{bmatrix} 1 \\ 1 \\ -a_R^+(\phi) \\ a_R^+(\phi) \end{bmatrix} \frac{e^{ik_F x - x/\xi}}{2\sqrt{\xi}}, \qquad (5.75)$$

$$\Phi_0^{(2)}(x) = \begin{bmatrix} 1 \\ -1 \\ a_R^-(2\pi - \phi) \\ a_R^-(2\pi - \phi) \end{bmatrix} \frac{e^{-ik_F x - x/\xi}}{2\sqrt{\xi}}, \qquad (5.76)$$

where $1/2\sqrt{\xi}$ is the normalization factor. Note that for the given energy (5.74) the Andreev amplitude a_R^- in $\Phi_0^{(2)}(x)$ has the phase $2\pi - \phi$. It is now easy to see that the states (5.75) and (5.76) are connected by the TRS operation:

$$\Phi_0^{(2)}(x) = \begin{bmatrix} i\sigma_y & 0 \\ 0 & i\sigma_y \end{bmatrix} \Phi_0^{(1)*}(x). \qquad (5.77)$$

Moreover, both particle and hole components of $\Phi_0^{(1,2)}(x)$ are the eigenstates of the spin projection σ_x, which can be expressed in the matrix form as

$$\begin{bmatrix} \sigma_x & 0 \\ 0 & -\sigma_x \end{bmatrix} \Psi_0^{(1,2)}(x) = \pm\Psi_0^{(1,2)}(x). \qquad (5.78)$$

Thus, the spins of these states are tied to the momentum direction along the x-axis. As a consequence of Eq. (5.77), the ABSs (5.75) and (5.76) are orthogonal to each other and therefore, protected against TRS-invariant perturbations.

Clearly, the protection by the TRS does not hold for the ABSs with oblique incidence at the potential barrier since in this case the incident and reflected trajectories are not related by the time reversal [see Fig. 5.8(b)]. The reflection from the potential barrier generates a minigap $\propto \sqrt{1-T_\theta}$ at $\phi = \pi, 3\pi, \ldots$, similar to the ABSs in conventional Josephson junctions [228].

Although for $\theta \neq 0$, the ABSs are not protected from potential scattering, they still feature the spin-momentum locking, and therefore, can be called helical in the same sense as the normal-state 2D surface carriers. It is worth noting that despite the dependence on the normal-state quantity T_θ the helical ABSs reflect the nature of the induced superconducting condensate, that is, its mixed s- and p-wave character. In Eq. (5.60) the s- and p-wave correlations are explicitly accounted for by the couplings between the opposite-spin (e.g., $u_\uparrow$ and $v_\downarrow$) and the same-spin (e.g., $u_\uparrow$ and $v_\uparrow$) particle and hole components, respectively. Alternatively, one can examine the spin structure of the Green function of the ABSs. This Green function can be constructed from the eigenspinors and energies of the ABSs by means of the Hilbert–Schmidt expansion. For our purpose it suffices to consider the contributions of the normally propagating states (5.75) and (5.76). For $E \to E_0^{(1)}(\phi)$, the contribution of the state (5.75) is obtained by making the direct product of the spinor $\Phi_0^{(1)}(x)$ with its conjugate $\tilde{\Phi}_0^{(1)}(x') = \left[1, 1, -a_R^{+*}, a_R^{+*}\right] \mathrm{e}^{-ik_F x' - x'/\xi}/(2\sqrt{\xi})$ as follows

$$G_0^{(1)}(x, x') \approx \frac{\Phi_0^{(1)}(x) \otimes \tilde{\Phi}_0^{(1)}(x')}{E - E_0^{(1)}(\phi)}$$

$$= \begin{bmatrix} 1 & 1 & -a_R^{+*} & a_R^{+*} \\ 1 & 1 & -a_R^{+*} & a_R^{+*} \\ -a_R^{+} & -a_R^{+} & 1 & -1 \\ a_R^{+} & a_R^{+} & -1 & 1 \end{bmatrix} \times \frac{\mathrm{e}^{ik_F(x-x') - (x+x')/\xi}}{4\xi\,[E - E_0^{(1)}(\phi)]} \tag{5.79}$$

$$= \begin{bmatrix} \sigma_0 + \sigma_x & (\sigma_0 + \sigma_x) i\sigma_y a_R^{+*} \\ -i\sigma_y(\sigma_0 + \sigma_x) a_R^{+} & \sigma_0 - \sigma_x \end{bmatrix} \times \frac{\mathrm{e}^{ik_F(x-x') - (x+x')/\xi}}{4\xi\,[E - E_0^{(1)}(\phi)]}.$$

The spin structure of the diagonal and off-diagonal matrix elements of Eq. (5.79) is similar to that of the Green functions in Eqs. (5.51) - (5.54) for $\boldsymbol{p}||x$. This proves that ABS (5.75) is of the mixed $s-$ and p-wave type. Similar calculation for the counter-propagating state (5.76) yields

$$G_0^{(2)}(x,x') \approx \frac{\Psi_0^{(2)}(x) \otimes \tilde{\Psi}_0^{(2)}(x')}{E - E_0^{(2)}(\phi)} \tag{5.80}$$

$$= \begin{bmatrix} \sigma_0 - \sigma_x & (\sigma_0 - \sigma_x) i\sigma_y a_R^+ \\ -i\sigma_y(\sigma_0 - \sigma_x) a_R^{+*} & \sigma_0 + \sigma_x \end{bmatrix} \times \frac{e^{-ik_F(x-x')-(x+x')/\xi}}{4\xi\,[E - E_0^{(2)}(\phi)]}.$$

The spin matrices in Eqs. (5.79) and (5.80) are related to each other by the TRS.

5.2.3 *Emergent Majorana Fermions*

We have seen that in S/TI/S junctions the TRS guaranties the existence of a pair of gapless ABSs regardless of the presence or absence of a potential barrier in the junction. We also know that there is another generic symmetry of the system—the particle-hole conjugation— which allows one to obtain any hole state by acting with the conjugation operator $\mathbb{C}$ (5.25) on the particle one or vice versa. This has an interesting implication for the states with an exactly zero energy. Due to the particle-hole symmetry, a zero-energy particle state must be identical to its hole counterpart and vice versa. Such zero-energy states are the condensed matter analogues of Majorana fermions of relativistic field theory, where those appear as particles and anti-particles at the same time. The physical origin of Majorana-like fermions in S/TI/S junctions and their properties are discussed in detail below (Ref. [209]).

Let us first understand why the zero-energy states appear only when the superconducting phase difference is an odd-integer of π,

$$\phi = \pi, 3\pi, \ldots \tag{5.81}$$

For simplicity, we consider the strictly one-dimensional junction in the x-direction, which is a particular case of the model presented in the preceding section. Under condition (5.81) the superconducting paring potential [see Eq. (5.58)] has the opposite signs in the left and the right S terminals:

$$\tilde{\Delta}|_L = \Delta_0, \qquad \tilde{\Delta}|_R = -\Delta_0. \tag{5.82}$$

The sign of $\tilde{\Delta}$ determines the relative ordering of the particle and hole spectral branches in a superconductor in the same manner

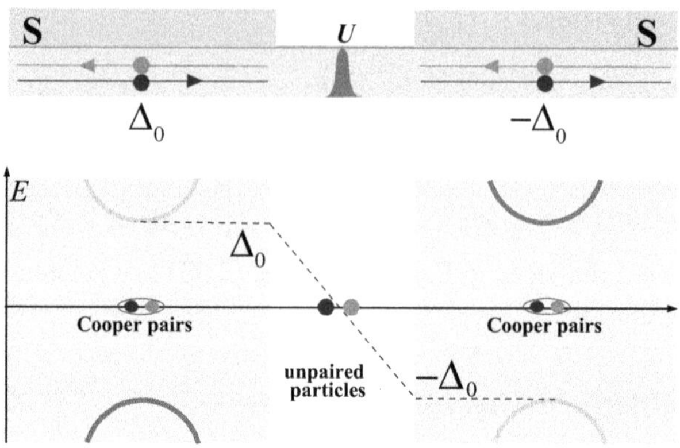

Figure 5.9 Schematic of the formation of zero-energy (midgap) states in an S/TI/S junction (see also text).

as the sign of the band gap determines the band sequence in an insulator. Therefore, under conditions (5.81) and (5.82) the ordering of the bands in one of the superconductors is inverted with respect to the other. This situation is illustrated in Fig. 5.9 where the particle (hole) branch corresponds to positive (negative) energies on the left and negative (positive) energies on the right. The inversion of the band structure in space implies that the gap between the particle and hole states must close in the middle of the junction. This has an important consequence. Due to the absence of the superconducting gap, there is no energy cost for adding unpaired particles to the ground state of the junction, whereas in the S terminals the ground state consists of Cooper pairs, and is separated from the excited single-particle states by a minimum energy of Δ_0.

The states accommodating the unpaired particles in the middle of the junction are indeed very special. First, they have zero energy with respect to the chemical potential and, for this reason, are related by the particle-hole conjugation. Second, the zero-energy states form a Kramers' doublet, and therefore, remain unaffected by a potential barrier in the junction. The particle-hole and TR symmetries impose the constraints on the zero-energy states, due to which the number of independent degrees of freedom is reduced to a pair of orthogonal Majorana fermions. To prove this let us take a closer look at the zero-energy solutions (5.75) and (5.76),

$$\Phi_0^{(1)}(x) = \begin{bmatrix} 1 \\ 1 \\ -i \\ i \end{bmatrix} \frac{e^{ik_F x - x/\xi}}{2\sqrt{\xi}}, \tag{5.83}$$

$$\Phi_0^{(2)}(x) = \begin{bmatrix} 1 \\ -1 \\ -i \\ -i \end{bmatrix} \frac{e^{-ik_F x - x/\xi}}{2\sqrt{\xi}}. \tag{5.84}$$

The two states transform into each other under the the particle-hole conjugation

$$\mathbb{C}\Phi_0^{(1)}(x) = \Phi_0^{(2)}(x), \qquad \mathbb{C}\Phi_0^{(2)}(x) = \Phi_0^{(1)}(x), \tag{5.85}$$

where the $\mathbb{C}$ operation is defined, for convenience, as

$$\mathbb{C} = \begin{bmatrix} 0 & -i\sigma_0 \\ -i\sigma_0 & 0 \end{bmatrix} C, \tag{5.86}$$

which differs from Eq. (5.24) only by the allowed phase factor $-i$. Let us now define two linear combinations of the zero-energy solutions as

$$\gamma_+(x) = \frac{\Phi_0^{(1)}(x) + \Phi_0^{(2)}(x)}{2}, \qquad \gamma_-(x) = \frac{\Phi_0^{(1)}(x) - \Phi_0^{(2)}(x)}{2i}. \tag{5.87}$$

From Eqs. (5.85) and (5.87) it is clear that $\gamma_\pm(x)$ are invariant under the charge conjugation:

$$\mathbb{C}\gamma_\pm(x) = \gamma_\pm(x). \tag{5.88}$$

In other words, $\gamma_\pm(x)$ are the particle and the hole states at the same time. In this sense, the self-conjugated solutions defined by Eq. (5.87) mimic the Majorana fermions of relativistic field theory where the role of $\mathbb{C}$ is played by the charge-conjugation operation of the Dirac equation. Importantly, in view of the TR relation $\Phi_0^{(2)}(x) = \mathbb{T}\Phi_0^{(1)}(x)$ between the zero-energy states, $\gamma_\pm(x)$ are orthogonal,

$$\langle\gamma_+|\gamma_-\rangle = 0, \tag{5.89}$$

and hence, independent states. We note that the expressions for $\gamma_\pm$ and $\mathbb{C}$ depend on the chosen basis, whereas the self-conjugation property (5.88) is generic. The emergent Majorana fermions are

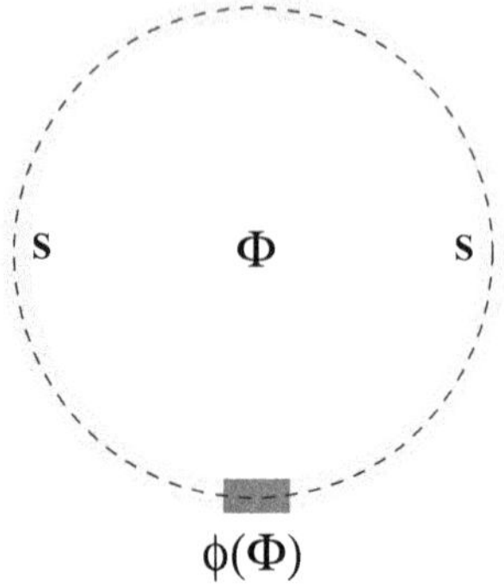

Figure 5.10 Scheme to control the Josephson phase difference ϕ at an S/TI/S junction in a superconducting ring with a magnetic flux Φ.

therefore a generic manifestation of the unconventional superconductivity in TIs.

In order to realize the "π-junction" with the zero-energy states, one should be able to control the Josephson phase difference ϕ. A way to achieve this is to incorporate an S/TI/S junction into a superconducting ring with a magnetic flux Φ, as depicted in Fig. 5.10. If the radius of the ring is much larger than its thickness, the phase difference depends on the magnetic flux through a simple formula

$$\phi(\Phi) = 2\pi \frac{\Phi}{\Phi_0}, \tag{5.90}$$

where $\Phi_0 = ch/2|e|$ is the magnetic flux quantum. Thus, for the half-integer magnetic flux $\Phi = \Phi_0/2, 3\Phi_0/2, \ldots$ the Josephson phase difference will be odd-integer of π.

Equation (5.90) can be derived from the principle of gauge invariance. The latter requires that in a magnetic field the local phase of the superconducting order parameter has the form

$$\phi(\boldsymbol{r}) = \phi_0(\boldsymbol{r}) - \frac{2e}{c\hbar} \int^{\boldsymbol{r}} \boldsymbol{A}(\boldsymbol{r}') \cdot d\boldsymbol{r}', \tag{5.91}$$

where $\phi_0(\boldsymbol{r})$ is the phase in the absence of the field, $\boldsymbol{A}(\boldsymbol{r})$ is the vector potential of the external magnetic field, and $2e$ is the charge of a Cooper pair. Let us take the phase gradient

$$\nabla\phi(\boldsymbol{r}) = \nabla\phi_0(\boldsymbol{r}) - \frac{2e}{c\hbar} \boldsymbol{A}(\boldsymbol{r}), \tag{5.92}$$

and integrate it along the ring in the anti-clockwise direction. Since the phase drop occurs mainly across the non-superconducting part of the junction, the integral of the left-hand side of Eq. (5.92) is $\phi_R - \phi_L = \phi$, whereas the right-hand side yields

$$\oint \nabla\phi_0(\boldsymbol{r}) \cdot d\boldsymbol{r} + 2\pi \frac{\Phi}{\Phi_0}. \tag{5.93}$$

For the uniquely defined order parameter the first term vanishes, which brings us to Eq. (5.90). Tuning the Josephson phase difference with the magnetic flux in a ring is one of the experimental tools to measure the Josephson current–phase relation.

5.2.4 *Josephson Effect. Superconducting Klein Tunneling*

The Josephson effect is a macroscopic quantum phenomenon that consists in the appearance of the electric current between two superconductors in the absence of the bias voltage. This current, called the supercurrent, is carried by Cooper pairs. The measurement of the supercurrent is one of the accessible tools to probe the symmetry of the superconducting order parameter. We may therefore ask the question of how the unconventional superconductivity and the topological bound states in TIs manifest themselves in the Josephson effect. In the context of 3DTIs, this question has been addressed in Refs. [56, 226, 227, 229].

In a short junction the supercurrent can be calculated from the spectrum of the ABSs according to the thermodynamic formula [228]:

$$I(\phi) = -\frac{eN}{\hbar} \int_0^{\pi/2} d\theta \cos\theta \, \frac{\partial E_\theta^+(\phi)}{\partial \phi} \tanh \frac{E_\theta^+(\phi)}{2\Xi}, \tag{5.94}$$

where $N = k_F W/\pi$ is the number of the transport channels in the system, Ξ is the temperature in energy units, and the integral in Eq. (5.94) accounts for all possible propagation directions in the junction plane. For the ABSs in Eq. (5.71) the current (5.94) is

$$\frac{I(\phi)}{I_0} = \frac{\sin\phi}{2} \int_0^{\pi/2} d\theta \cos\theta \frac{T_\theta}{\left[1 - T_\theta \sin^2 \frac{\phi}{2}\right]^{1/2}} \tanh \frac{\left[1 - T_\theta \sin^2 \frac{\phi}{2}\right]^{1/2}}{2\Xi/\Delta_0} \tag{5.95}$$

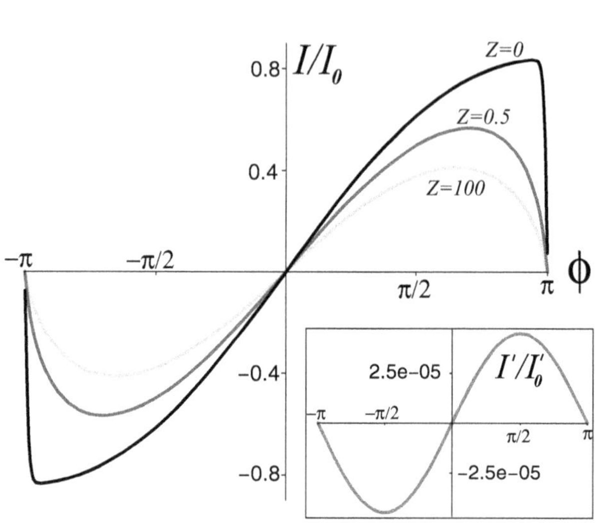

Figure 5.11 Current-phase relation $I(\phi)$ (5.95) in S/TI/S junction for different barrier strengths Z. Inset: $I'(\phi)$ (5.96) for a conventional reference S/N/S junction for strong barrier with $Z = 100$. In both cases $\Xi = 0.01\Delta_0$ (adapted from Ref. [227]).

where $I_0 = eN\Delta_0/2\hbar$. It is instructive to consider in parallel a reference S/N/S junction, where N is a conventional 2D system with a spin-degenerate parabolic dispersion (5.26). We assume that the reference system has the same geometry and the same (delta-like) potential in the middle. In this case, the ABSs are also given by Eq. (5.71), but with the angle-independent transmission probability $T' = 1/(1 + Z^2)$ for the non-helical carriers in the N. Consequently, for the reference supercurrent, $I'(\phi)$, Eq. (5.94) gives

$$\frac{I'(\phi)}{I'_0} = \frac{\sin\phi}{2} \frac{T'}{\left[1 - T' \sin^2 \frac{\phi}{2}\right]^{1/2}} \tanh \frac{\left[1 - T' \sin^2 \frac{\phi}{2}\right]^{1/2}}{2\Xi/\Delta_0}, \quad (5.96)$$

where $I'_0 = 2I_0$ includes the factor of two due to the spin degeneracy.

The equilibrium current–phase relation (5.95) is 2π-periodic. It is plotted in Fig. 5.11 for different values of the barrier strengths Z. The most interesting behavior is observed in the case of large Z. In this case, the conventional current–phase relation (5.96) is sinusoidal,

$$I'(\phi) \propto \sin\phi, \quad (5.97)$$

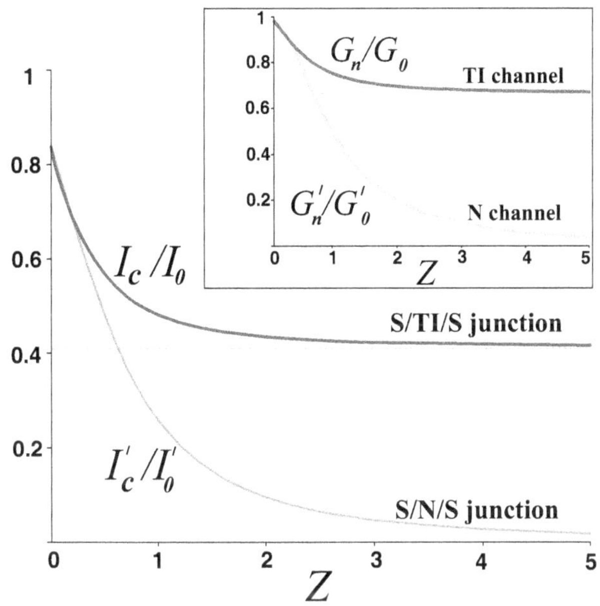

Figure 5.12 Normalized critical current versus barrier strength Z for S/TI/S and conventional S/N/S junctions. Inset: Normal-state conductance versus barrier strength Z for TI and N channels (see also text); $\Xi = 0.01\Delta_0$ (adapted from Ref. [227]).

as shown for comparison in the inset of Fig. 5.11. In sharp contrast to Eq. (5.97) for the S/TI/S the current $I(\phi)$ remains non-sinusoidal and is by orders of magnitude larger than the conventional current $I'(\phi)$. The reason for this drastic difference is the gapless ABSs (5.71) whose contribution is $\propto \sin(\phi/2)\mathrm{sgn}(\cos(\phi/2))$ and independent of the barrier strength. Thus, the first signature of the unconventional superconductivity and the topological bound states in TIs is the non-sinusoidal Josephson current–phase relation.

Next, we analyze numerically the critical supercurrent I_c. It is defined as the maximum of $I(\phi)$ (5.95) at fixed parameters Z and Ξ. Figure 5.12 shows the normalized I_c as a function of the barrier strength Z. Initially decreasing with Z, the critical current saturates at a constant value for $Z > 3$. This is again in sharp contrast to the conventional S/N/S junction in which $I'_c(Z)$ is strongly suppressed for $Z > 3$. The large supercurrent despite the strong barrier is a

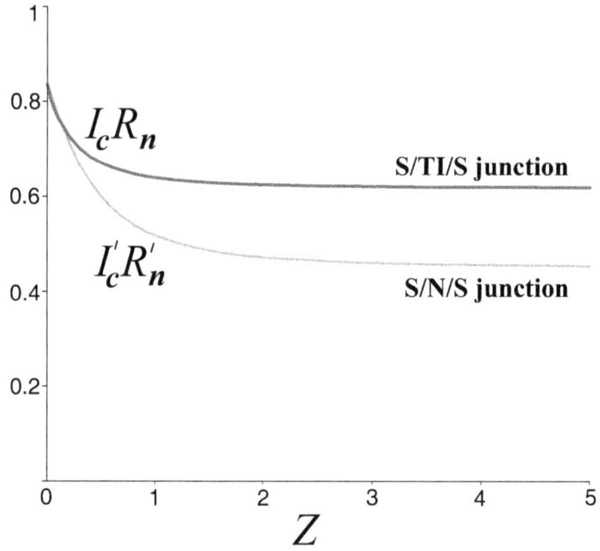

Figure 5.13 $I_c R_n$ product in units of $\pi\Delta_0/e$ versus barrier strength Z for S/TI/S and conventional S/N/S junctions; $\Xi = 0.01\Gamma_0$ (adapted from Ref. [227]).

manifestation of the helical ABSs (5.71) which are able to transport the Cooper pairs through the barrier without reflection for $\theta = 0$. This is a superconducting analogue of the normal-state Klein tunneling. For comparison, the inset in Fig. 5.12 shows the normal-state conductance of the TI:

$$G_n = G_0 \int_0^{\pi/2} d\theta \cos\theta \, T_\theta, \quad G_0 = \frac{e^2 N}{h}, \tag{5.98}$$

along with that of the conventional N system $G_n' = G_0' T'$, with $G_0' = 2G_0$. Klein tunneling with $T_0 = 1$ for $\theta = 0$ leads to the saturation of the conductance G_n at large Z, which clearly correlates with the behavior of the critical current I_c. Figure 5.13 shows another experimentally relevant characteristic, the product of the critical current I_c and the normal-state resistance $R_n = 1/G_n$ (5.98) normalized by $\pi\Delta_0/\pi$. Due to the Klein tunneling, the $I_c R_n$ product is larger for the S/TI/S junction than for the conventional S/N/S system.

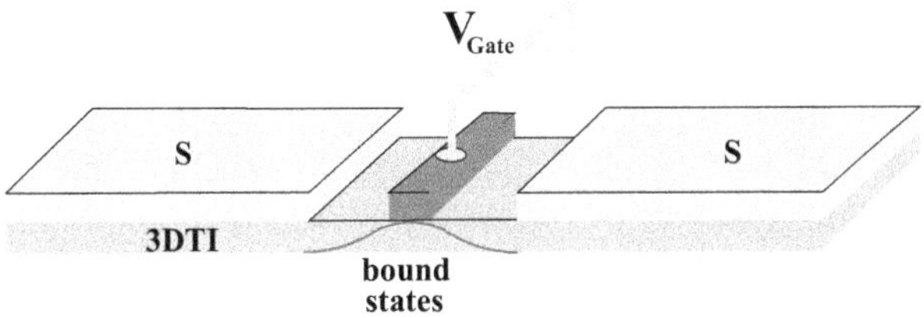

Figure 5.14 Schematic of a gated Josephson junction device for all-electrical detection of the topological bound states through superconducting Klein tunneling.

Summarizing the results of this section, we can conclude that the Josephson current–phase relation, critical current and the $I_c R_n$ product all show the signs of the reflectionless passage of Cooper pairs through the junction. This superconducting Klein tunneling effect is directly linked to the topological bound states and, therefore, can be used to detect them. The detection scheme requires a tunable potential barrier for electrons in the normal part of the junction. Such a barrier can be formed by electrostatic gating, as depicted in Fig. 5.14. This all-electrical detection scheme should be realizable already in the present-day S/TI/S Josephson devices.

5.2.5 *Related Research on Topological Superconductivity*

We have considered only one example of topological Josephson junctions, which does not exhaust all the interesting and experimentally relevant situations. Topological bound states have been studied theoretically in a variety of superconducting TI structures (see, e.g., Refs. [230–241]). Among the proposals to detect the gapless ABSs and the emergent Majorana states are the 4π-periodic Josephson effect [36, 239, 240] and its signatures in the current noise [236], quantum interferometry [231], resonant Andreev reflection [232], magnetic proximity effect [233, 234], anomalies in the current–phase relation and the Fraunhofer pattern [241], to name the few. Besides, alternative condensed matter platforms have been considered for engineering the emergent Majorana fermions. One example is the hybrid structures of superconductors and semiconductor quantum wires with Rashba spin–orbit coupling and strong Zeeman splitting [242–245].

The unambiguous observation of Majorana fermions still remains an outstanding goal. The pursuit of Majorana fermions in solid state systems is motivated by the exotic properties of these quasiparticles; the nonAbelian exchange statistics that they generate and their potential for topological quantum computation (see reviews in Refs. [18, 19, 246, 247]).

5.3 Problems

(1) Consider the Hamiltonian

$$H = \int d\boldsymbol{r} \sum_{\alpha,\beta=\uparrow,\downarrow} \Big[\psi^{\dagger}_{\alpha}(\boldsymbol{r}, t) h_{\alpha\beta}(\boldsymbol{r}) \psi_{\beta}(\boldsymbol{r}, t) + \frac{1}{2}\psi^{\dagger}_{\alpha}(\boldsymbol{r}, t)\Delta_{\alpha\beta}\psi^{\dagger}_{\beta}(\boldsymbol{r}, t) + \frac{1}{2}\psi_{\alpha}(\boldsymbol{r}, t)[\Delta_{\alpha\beta}]^{\dagger}\psi_{\beta}(\boldsymbol{r}, t)\Big], \tag{5.99}$$

where $[\Delta_{\alpha\beta}]^{\dagger}$ denotes the hermitian conjugation

$$[\Delta_{\alpha\beta}]^{\dagger} = \Delta^{*}_{\beta\alpha}. \tag{5.100}$$

Derive the equations of motion for the operators $\psi_{\alpha}(\boldsymbol{r}, t)$ and $\psi^{\dagger}_{\alpha}(\boldsymbol{r}, t)$. Prove that they coincide with Eqs. (5.8) and (5.9) provided that $\Delta_{\alpha\beta}$ is antisymmetric under the exchange of the spin indices: $\Delta_{\alpha\beta} = -\Delta_{\beta\alpha}$. Equation (5.99) is therefore, the mean-field Hamiltonian of a spin singlet superconductor.

(2) Using the results of Section 5.1.2, calculate the density of states (DOS)

$$N_S(E) = \sum_{s\boldsymbol{p}} \delta(E - E_{s\boldsymbol{p}}) \tag{5.101}$$

for a conventional superconductor with the $s-$wave pairing and spin-independent quadratic normal-state spectrum. Show that for a large Fermi energy $\mu \gg |\Delta|$ the DOS is given by

$$N_S(E) = N\frac{|E|}{\sqrt{E^2 - |\Delta|^2}}\Theta(|E| - |\Delta|), \tag{5.102}$$

where N is the normal-state DOS at the Fermi level.

(3) Calculate the Nambu Green function of a conventional superconductor from the matrix equation

$$\left\{ E\hat{I} - \begin{bmatrix} \frac{p^2}{2m} - \mu & i\sigma_y\tilde{\Delta} \\ -i\sigma_y\tilde{\Delta}^* & -\frac{p^2}{2m} + \mu \end{bmatrix} \right\} \begin{bmatrix} \hat{G}_{11,\boldsymbol{p}E} & \hat{G}_{12,\boldsymbol{p}E} \\ \hat{G}_{21,\boldsymbol{p}E} & \hat{G}_{22,\boldsymbol{p}E} \end{bmatrix} = \hat{I}. \quad (5.103)$$

Show that for a large Fermi energy $\mu \gg |\Delta|$ the Green function is given by

$$\hat{G}_{\boldsymbol{p}E} = \begin{bmatrix} \hat{G}_{11,\boldsymbol{p}E} & \hat{G}_{12,\boldsymbol{p}E} \\ \hat{G}_{21,\boldsymbol{p}E} & \hat{G}_{22,\boldsymbol{p}E} \end{bmatrix} = \frac{\begin{bmatrix} \sigma_0\left(E + \frac{p^2}{2m} - \mu\right) & i\sigma_y\,\Delta \\ -i\sigma_y\,\Delta^* & \sigma_0\left(E - \frac{p^2}{2m} + \mu\right) \end{bmatrix}}{E^2 - \left(\frac{p^2}{2m} - \mu\right)^2 - |\Delta|^2}. \quad (5.104)$$

(4) The knowledge of the Green function allows us to calculate various thermodynamic characteristics of the superconductor. For example, its DOS can be related to the imaginary part of the quasiparticle Green function $\hat{G}_{11,\boldsymbol{p}E}$ as

$$N_S(E) = -\frac{1}{\pi}\sum_{\boldsymbol{p}} \mathrm{Tr}(\Im\hat{G}_{11,\boldsymbol{p}E+i0}), \quad (5.105)$$

where Tr denotes the trace operation over the spin degree of freedom. Changing the integration variable from $\boldsymbol{p}$ to $\xi_{\boldsymbol{p}} = \mathbf{p}^2/2m - \mu$ as described in Section 4.4.3, derive Eq. (5.102) for the DOS.

(5) The BdG Hamiltonian (5.13) can be used to describe superconductors with broken TRS. An illustrative example is a superconductor carrying a dissipationless current. For the particles with a spin-independent quadratic spectrum the current-carrying superconducting state is described by Eq. (5.13) in which the Hamiltonian $\hat{h}$ is given by

$$\hat{h}(\boldsymbol{r}) = \left[\frac{(-i\hbar\nabla + \boldsymbol{q}/2)^2}{2m} - \mu\right]\sigma_0, \quad (5.106)$$

where the vector $\boldsymbol{q}$ has the meaning of the momentum of the Cooper-pair condensate. Write the corresponding BdG equations in the momentum representation [*cf.* Eqs. (5.27) and (5.28)]. Assuming $\mu \gg \boldsymbol{q}^2/8m$, prove that the excitation spectrum of the superconductor is

$$E_{s\boldsymbol{p}} = -\frac{\boldsymbol{p}\cdot\boldsymbol{q}}{2m} \pm \sqrt{\left(\frac{\boldsymbol{p}^2}{2m} - \mu\right)^2 + |\Delta|^2}. \qquad (5.107)$$

What happens when $\frac{|\boldsymbol{p}\cdot\boldsymbol{q}|}{2m} = |\Delta|$?

(6) Starting from the expression (5.38) for the condensate correlation function, obtain the classification of the superconducting pairing in the Nambu basis defined in Eq. (5.17). What are the spin functions of the singlet and triplet Cooper pairs in this basis?

Appendix A

Weak Antilocalization in Topological Insulators

A.1 Equations (4.137) for Cooperon amplitudes

In this appendix we outline the derivation of Eq. (4.137) for the Cooperon amplitudes $C^{ij}(\omega, \boldsymbol{q})$. Inverting the expansion in Eq. (4.135), we have

$$C^{ij}(\omega, \boldsymbol{q}) = \sum_{\alpha\beta\alpha'\beta'} \Psi^{i*}_{\alpha\beta} \Psi^{j}_{\alpha'\beta'} C_{\alpha\beta\alpha'\beta'}(\omega, \boldsymbol{q}). \tag{A.1}$$

Inserting this into Eq. (4.134) yields the equation

$$C^{ij}(\omega, \boldsymbol{q}) = \frac{\tau^2}{\tau_0}\delta_{ij} + \zeta \sum_{s} \int \frac{d\boldsymbol{k}}{(2\pi)^2} \tag{A.2}$$

$$\times \left[\sum_{\alpha\beta\gamma'\delta'} \Psi^{i*}_{\alpha\beta} G^{R}_{\alpha\gamma'}(\boldsymbol{k}, E + \hbar\omega) G^{A}_{\beta\delta'}(\boldsymbol{q} - \boldsymbol{k}, E) \Psi^{s}_{\gamma'\delta'} \right] C^{sj}(\omega, \boldsymbol{q}).$$

It can be rewritten in a simpler form with the help of the identity

$$\sum_{\alpha\beta\gamma'\delta'} \Psi^{i*}_{\alpha\beta} G^{R}_{\alpha\gamma'}(\boldsymbol{k}, E + \hbar\omega) G^{A}_{\beta\delta'}(\boldsymbol{q} - \boldsymbol{k}, E) \Psi^{s}_{\gamma'\delta'}$$

$$= \sum_{\delta'\beta\alpha\gamma'} \left[G^{A^T}(\boldsymbol{q} - \boldsymbol{k}, E) \right]_{\delta'\beta} \left[\Psi^{i\dagger} \right]_{\beta\alpha} G^{R}_{\alpha\gamma'}(\boldsymbol{k}, E + \hbar\omega) \Psi^{s}_{\gamma'\delta'}$$

$$= \mathrm{Tr}\left[\hat{G}^{A^T}(\boldsymbol{q} - \boldsymbol{k}, E) \Psi^{i\dagger} \hat{G}^{R}(\boldsymbol{k}, E + \hbar\omega) \Psi^{s} \right], \tag{A.3}$$

where Tr and T denote the trace and transposition operations, respectively, in $\boldsymbol{\sigma}$ space. We therefore have

$$C^{ij}(\omega, \boldsymbol{q}) = \frac{\tau^2}{\tau_0}\delta_{ij} + \zeta \sum_{s} \mathrm{Tr}\left[\int \frac{d\boldsymbol{k}}{(2\pi)^2} \right.$$

$$\left. \times \hat{G}^{A^T}(\boldsymbol{q} - \boldsymbol{k}, E) \Psi^{i\dagger} \hat{G}^{R}(\boldsymbol{k}, E + \hbar\omega) \Psi^{s} \right] C^{sj}(\omega, \boldsymbol{q}). \tag{A.4}$$

For the chosen orthonormal spin basis $\Psi^j = \sigma_j\sigma_y/\sqrt{2}$ [see also Eq. (4.136)] Eq. (A.4) assumes the following form:

$$C^{ij}(\omega, \boldsymbol{q}) = \frac{\tau^2}{\tau_0}\delta_{ij} + \frac{\zeta}{2}\sum_s \mathrm{Tr}\Big[\int \frac{d\boldsymbol{k}}{(2\pi)^2}$$

$$\times\widetilde{\hat{G}^A}(\boldsymbol{q}-\boldsymbol{k}, E)\sigma_i\hat{G}^R(\boldsymbol{k}, E+\hbar\omega)\sigma_s\Big]C^{sj}(\omega, \boldsymbol{q}). \tag{A.5}$$

Here the tilde denotes the time-reversal operation,

$$\widetilde{\hat{G}^A} = \sigma_y\left[\hat{G}^A\right]^T\sigma_y, \tag{A.6}$$

which flips the spin: $\widetilde{\boldsymbol{\sigma}} = \sigma_y\boldsymbol{\sigma}^T\sigma_y = -\boldsymbol{\sigma}$. As expected, for $\boldsymbol{q} \to 0$ Eq. (A.5) describes interference between the state with $\boldsymbol{k}, \boldsymbol{\sigma}$, and its "time-reversed" partner with $-\boldsymbol{k}, -\boldsymbol{\sigma}$.

Next, we evaluate the $\boldsymbol{k}$ integral in Eq. (A.5). For that we again use the sharpness of the Green functions (4.123) near the Fermi surface, $|\boldsymbol{k}| \approx k_F$, under condition (4.81) as well as the smallness of the Cooperon momentum

$$|\boldsymbol{q}| \ll k_F. \tag{A.7}$$

We expand the denominator of $\widetilde{\hat{G}^A}_{\boldsymbol{q}-\boldsymbol{k}}$ [see Eq. (4.123)] up to the linear order in $\boldsymbol{q}$, that is, $\xi_{\boldsymbol{q}-\boldsymbol{k}} \approx \xi_{\boldsymbol{k}} - \hbar v\boldsymbol{n}\cdot\boldsymbol{q}$ (the last term should be compared with $\hbar/\tau$), which yields

$$\widetilde{\hat{G}^A}_{\boldsymbol{q}-\boldsymbol{k}} \approx \frac{\widetilde{P}_{-\boldsymbol{n}}}{E_A - \xi_{\boldsymbol{k}} + \hbar v\boldsymbol{n}\cdot\boldsymbol{q}}. \tag{A.8}$$

With Eqs. (4.123) and (A.8), the $\boldsymbol{k}$ integral can be evaluated as follows

$$\int \frac{d\boldsymbol{k}}{(2\pi)^2}\widetilde{G^A}(\boldsymbol{q}-\boldsymbol{k}, E)\sigma_i G^R(\boldsymbol{k}, E+\hbar\omega)\sigma_s \tag{A.9}$$

$$= \int \frac{d\phi_{\boldsymbol{n}}}{2\pi}\int_{-E_F}^{\infty} N(\xi, \boldsymbol{n})d\xi\frac{\widetilde{P}_{-\boldsymbol{n}}\sigma_i P_{\boldsymbol{n}}\sigma_s}{(E_A - \xi + \hbar v\boldsymbol{n}\cdot\boldsymbol{q})(E_R - \xi)} \tag{A.10}$$

$$\approx N\int \frac{d\phi_{\boldsymbol{n}}}{2\pi}\widetilde{P}_{-\boldsymbol{n}}\sigma_i P_{\boldsymbol{n}}\sigma_s\int_{-\infty}^{\infty}\frac{d\xi}{(\xi - E_A - \hbar v\boldsymbol{n}\cdot\boldsymbol{q})(\xi - E_R)} \tag{A.11}$$

$$= N\int \frac{d\phi_{\boldsymbol{n}}}{2\pi}\widetilde{P}_{-\boldsymbol{n}}\sigma_i P_{\boldsymbol{n}}\sigma_s\frac{2\pi i}{i\hbar/\tau + \hbar\omega - \hbar v\boldsymbol{n}\cdot\boldsymbol{q}} \tag{A.12}$$

$$= \frac{2\pi N\tau}{\hbar}\int \frac{d\phi_{\boldsymbol{n}}}{2\pi}\frac{\widetilde{P}_{-\boldsymbol{n}}\sigma_i P_{\boldsymbol{n}}\sigma_s}{1 - i\tau\omega + i\tau v\boldsymbol{n}\cdot\boldsymbol{q}}. \tag{A.13}$$

In Eq. (A.11) the exact DOS $N(\xi, \boldsymbol{n})$ is replaced by its Fermi surface value N and the lower ξ-integration limit is extended from $-E_F$ to $-\infty$ under condition (4.81). Then, the ξ integral is calculated along the contour encircling the upper complex half-plane with a pole at

$$\xi = E_R = E + \hbar\omega + \frac{i\hbar}{2\tau}. \tag{A.14}$$

Inserting Eq. (A.12) into Eq. (A.5) and using Eq. (4.92) for ζ, we arrive at Eq. (4.137).

A.2 Evaluation of *k* integrals in Hikami boxes in Eqs. (4.167) and (4.168)

To calculate the $\boldsymbol{k}$ integral in Eq. (4.167) we first expand the Green functions $\hat{G}^R_{\boldsymbol{q}-\boldsymbol{k},E+\hbar\omega}$, and $\hat{G}^A_{\boldsymbol{q}-\boldsymbol{k},E}$ in small Cooperon momentum $\boldsymbol{q}$ as we did in Eq. (A.8) of Appendix A.1:

$$\hat{G}^R_{\boldsymbol{q}-\boldsymbol{k},E+\hbar\omega} \approx \frac{\hat{P}_{-\boldsymbol{n}}}{E_R - \xi_{\boldsymbol{k}} + \hbar\omega + \hbar v \boldsymbol{n}\cdot\boldsymbol{q}}, \tag{A.15}$$

$$\hat{G}^A_{\boldsymbol{q}-\boldsymbol{k},E} \approx \frac{\hat{P}_{-\boldsymbol{n}}}{E_A - \xi_{\boldsymbol{k}} + \hbar v \boldsymbol{n}\cdot\boldsymbol{q}} \tag{A.16}$$

Then, the integral is converted to $\int \frac{d\phi_{\boldsymbol{n}}}{2\pi} \int N d\xi \ldots$ and the ξ-integration is done in the complex plane following the same steps as in Eqs. (A.9)–(A.12). The final result is

$$\int \frac{d\boldsymbol{k}}{(2\pi)^2} (\hat{G}^A_{\boldsymbol{k},E} \hat{\mathcal{V}}^x_{\boldsymbol{k}} \hat{G}^R_{\boldsymbol{k},E+\hbar\omega})_{\gamma'\beta} (\hat{G}^R_{\boldsymbol{q}-\boldsymbol{k},E+\hbar\omega} \hat{\mathcal{V}}^x_{\boldsymbol{q}-\boldsymbol{k}} \hat{G}^A_{\boldsymbol{q}-\boldsymbol{k},E})_{\gamma\beta'}$$
$$\approx \frac{4\pi N\tau^3}{\hbar^3} \overline{[\hat{P}_{\boldsymbol{n}} \hat{\mathcal{V}}^x_{\boldsymbol{n}} \hat{P}_{\boldsymbol{n}}]_{\gamma'\beta} [\hat{P}_{-\boldsymbol{n}} \hat{\mathcal{V}}^x_{-\boldsymbol{n}} \hat{P}_{-\boldsymbol{n}}]_{\gamma\beta'}}, \tag{A.17}$$

where $\overline{(\ldots)} = \int_0^{2\pi} \ldots d\phi_{\boldsymbol{n}}/2\pi$ is the averaging over the momentum direction $\boldsymbol{n}$. The equation above is independent of E, so that the energy integral in Eq. (4.167) is $\hbar/2\pi$, which along with Eq. (A.17) yields Eq. (4.169).

To evaluate the integrals over $\boldsymbol{k}$ and $\boldsymbol{k}'$ in Eq. (4.168) we set $\boldsymbol{q}, \omega \to 0$ [see Eq. (4.142)] in the Green functions and re-group them as follows

$$\int \frac{d\boldsymbol{k}}{(2\pi)^2} \int \frac{d\boldsymbol{k}'}{(2\pi)^2} \ (\hat{G}^A_{\boldsymbol{k},E} \ \hat{\mathcal{V}}^x_{\boldsymbol{k}} \ \hat{G}^R_{\boldsymbol{k},E+\hbar\omega} \hat{G}^R_{\boldsymbol{q}-\boldsymbol{k}',E+\hbar\omega})_{\gamma'\beta} \tag{A.18}$$

$$\times (\hat{G}^R_{\boldsymbol{q}-\boldsymbol{k},E+\hbar\omega} \ \hat{G}^R_{\boldsymbol{k}',E+\hbar\omega} \ \hat{\mathcal{V}}^x_{\boldsymbol{k}'} \ \hat{G}^A_{\boldsymbol{k}',E})_{\gamma\beta'}$$

$$\approx \int \frac{d\boldsymbol{k}}{(2\pi)^2}[\hat{G}^A(\boldsymbol{k})\hat{\mathcal{V}}^x(\boldsymbol{k})\hat{G}^R(\boldsymbol{k})]_{\gamma'\beta_1}\hat{G}^R_{\gamma\gamma_1}(-\boldsymbol{k})$$
$$\times \int \frac{d\boldsymbol{k}'}{(2\pi)^2}\hat{G}^R_{\beta_1\beta}(-\boldsymbol{k}')[\hat{G}^R(\boldsymbol{k}')\hat{\mathcal{V}}^x(\boldsymbol{k}')\hat{G}^A(\boldsymbol{k}')]_{\gamma_1\beta'}. \tag{A.19}$$

Each integral can now be done in the similar way as in Eqs. (A.9)–(A.12) of Appendix A.1:

$$\int \frac{d\boldsymbol{k}}{(2\pi)^2}[\hat{G}^A(\boldsymbol{k})\hat{\mathcal{V}}^x(\boldsymbol{k})\hat{G}^R(\boldsymbol{k})]_{\gamma'\beta_1}\hat{G}^R_{\gamma\gamma_1}(-\boldsymbol{k})$$
$$\approx -\frac{2\pi i N\tau^2}{\hbar^2}\overline{[\hat{P}_{\boldsymbol{n}}\hat{\mathcal{V}}^x_{\boldsymbol{n}}\hat{P}_{\boldsymbol{n}}]_{\gamma'\beta_1}[\hat{P}_{-\boldsymbol{n}}]_{\gamma\gamma_1}}, \tag{A.20}$$

$$\int \frac{d\boldsymbol{k}'}{(2\pi)^2}\hat{G}^R_{\beta_1\beta}(-\boldsymbol{k}')[\hat{G}^R(\boldsymbol{k}')\hat{\mathcal{V}}^x(\boldsymbol{k}')\hat{G}^A(\boldsymbol{k}')]_{\gamma_1\beta'}$$
$$\approx -\frac{2\pi i N\tau^2}{\hbar^2}\overline{[\hat{P}_{-\boldsymbol{n}'}]_{\beta_1\beta}[\hat{P}_{\boldsymbol{n}'}\hat{\mathcal{V}}^x_{\boldsymbol{n}'}\hat{P}_{\boldsymbol{n}'}]_{\gamma_1\beta'}}. \tag{A.21}$$

Inserting these into Eq. (4.168) we obtain Eq. (4.170).

Bibliography

1. Kane, C. L. and Mele, E. J. (2005), Quantum spin Hall effect in graphene, *Phys. Rev. Lett.* **95**, 226801.
2. Kane, C. L. and Mele, E. J. (2005), Z_2 topological order and the quantum spin Hall effect, *Phys. Rev. Lett.* **95**, 146802.
3. Bernevig, B. A. and Zhang, S.-C. (2006), Quantum spin Hall effect, *Phys. Rev. Lett.* **96**, 106802.
4. Bernevig, B. A., Hughes, T. L. and Zhang, S. C. (2006), Quantum spin Hall effect and topological phase transition in HgTe quantum wells, *Science* **314**, 1757–1761.
5. König, M., Wiedmann, S., Brüne, C., Roth, A., Buhmann, H., Molenkamp, L. M., Qi, X.-L. and Zhang, S.-C. (2007), Quantum spin Hall insulator state in HgTe quantum wells, *Science* **318**, 766–770.
6. König, M., Buhmann, H., Molenkamp, L. W., Hughes, T., Liu, C.-X., Qi, X.-L. and Zhang, S.-C. (2008), The quantum spin Hall effect: Theory and experiment, *J. Phys. Soc. Jpn.* **77**, 031007.
7. Roth, A., Brüne, C. Buhmann, H., Molenkamp, L. W., Maciejko, J., Qi, X.-L. and Zhang, S.-C. (2009), Nonlocal transport in the quantum spin Hall state, *Science* **325**, 294–297.
8. Fu, L., Kane, C. L. and Mele, E. J. (2007), Topological insulators in three dimensions, *Phys. Rev. Lett.* **98**, 106803.
9. Fu, L. and Kane, C. L. (2007), Topological insulators with inversion symmetry, *Phys. Rev. B* **76**, 045302.
10. Moore, J. E. and Balents, L. (2007), Topological invariants of time-reversal-invariant band structures, *Phys. Rev. B* **75**, 121306(R).
11. Hsieh, D., Qian, D., Wray, L., Xia, Y., Hor, Y. S., Cava, R. J. and Hasan, M. Z. (2008), A topological Dirac insulator in a quantum spin Hall phase, *Nature* **452**, 970–974.
12. Hsieh, D., Xia, Y., Wray, L., Qian, D., Pal, A., Dil, J. H., Meier, F., Osterwalder, J., Kane, C. L., Bihlmayer, G., Hor, Y. S., Cava, R. J. and Hasan, M. Z. (2009),

Observation of unconventional quantum spin textures in topological insulators, *Science* **323**, 919–922.

13. Wigner, E. P. (1931), *Gruppentheorie* (Friedrich Vieweg und Sohn, Braunschweig, Germany), pp. 251–254; (1959), *Group Theory* (Academic Press Inc., New York), pp. 233–236.
14. Halperin, B. I. (1982), Quantized Hall conductance, current-carrying edge states, and the existence of extended states in a two-dimensional disordered potential, *Phys. Rev. B* **25**, 2185–2190.
15. MacDonald, A. H. and Streda, P. (1984), Quantized Hall effect and edge currents, *Phys. Rev. B* **29**, 1616.
16. v. Klitzing, K., Dorda, G. and Pepper, M. (1980), New method for high-accuracy determination of the fine-structure constant based on quantized Hall resistance, *Phys. Rev. Lett.* **45**, 494–497.
17. Anderson, P. W. (1958), Absence of diffusion in certain random lattices, *Phys. Rev.* **109**, 1492–1505.
18. Hasan, M. Z. and Kane, C. L. (2010), Colloquium: Topological insulators, *Rev. Mod. Phys.* **82**, 3045–3067.
19. Qi, X.-L. and Zhang, S.-C. (2011), Topological insulators and superconductors, *Rev. Mod. Phys.* **83**, 1057–1110.
20. Volkov, B. A. and Pankratov, O. A. (1985), Two-dimensional massless electrons in an inverted contact, *JETP Lett.* **42**, 178–181.
21. Pankratov, O. A., Pakhomov, S. V. and Volkov, B. A. (1987), Supersymmetry in heterojunctions: Band-inverting contact on the basis of $Pb_{1-x}Sn_xTe$ and $Hg_{1-x}Cd_xTe$, *Sol. State Commun.* **61**, 93.
22. Zhang, H., Liu, C.-X., Qi, X.-L., Dai, X., Fang, Z. and Zhang, S.-C. (2009), Topological insulators in Bi_2Se_3, Bi_2Te_3 and Sb_2Te_3 with a single Dirac cone on the surface, *Nat. Phys.* **5**, 438–442.
23. Yan, B., Liu, C.-X., Zhang, H.-J., Yam, C.-Y., Qi, X.-L., Frauenheim, T. and Zhang, S.-C. (2010), Theoretical prediction of topological insulators in thallium-based III-V-VI2 ternary chalcogenides, *EPL* **90**, 37002.
24. Lin, H., Markiewicz, R. S., Wray, L. A., Fu, L., Hasan, M. Z. and Bansil, A. (2010), Single-Dirac-cone topological surface states in the $TlBiSe_2$ class of topological semiconductors, *Phys. Rev. Lett.* **105**, 036404.
25. Eremeev, S. V., Koroteev, Yu. M. and Chulkov, E. V. (2010), Ternary thallium-based semimetal chalcogenides Tl-V-VI2 as a new class of three-dimensional topological insulators, *Pis'ma Zh. Eksp. Teor. Fiz.* **91**, 664; (2010) [*JETP Lett.* **91**, 594].

26. Eremeev, S. V., Koroteev, Yu. M. and Chulkov, E. V. (2010), On possible deep subsurface states in topological insulators: The $PbBi_4Te_7$ system, *Pis'ma Zh. Eksp. Teor. Fiz.* **92**, 183; [*JETP Lett.* **92**, 161].

27. Jin, H., Song, J. H, Freeman, A. J. and Kanatzidis, M. G. (2011), Candidates for topological insulators: Pb-based chalcogenide series, *Phys. Rev. B* **83**, 041202.

28. Xia, Y., Qian, D., Hsieh, D., Wray, L., Pal, A., Lin, H., Bansil, A., Grauer, D., Hor, Y. S., Cava, R. J. and Hasan, M. Z. (2009), Observation of a large-gap topological-insulator class with a single Dirac cone on the surface, *Nat. Phys.* **5**, 398–402.

29. Chen, Y. L., Analytis, J. G., Chu, J.-H., Liu, Z. K., Mo, S.-K., Qi, X. L., Zhang, H. J., Lu, D. H., Dai, X., Fang, Z., Zhang, S. C., Fisher, I. R., Hussain, Z. and Shen, Z.-X. (2009), Experimental realization of a three-dimensional topological insulator, Bi_2Te_3, *Science* **325**, 178–181.

30. Sato, T., Segawa, K., Guo, H., Sugawara, K., Souma, S., Takahashi, T. and Ando, Y. (2010), Direct evidence for the Dirac-cone topological surface states in the ternary chalcogenide $TlBiSe_2$, *Phys. Rev. Lett.* **105**, 136802.

31. Kuroda, K., Ye, M., Kimura, A., Eremeev, S. V., Krasovskii, E. E., Chulkov, E. V., Ueda, Y., Miyamoto, K., Okuda, T., Shimada, K., Namatame, H. and Taniguchi, M. (2010), Experimental realization of a three-dimensional topological insulator phase in ternary chalcogenide $TlBiSe_2$, *Phys. Rev. Lett.* **105**, 146801.

32. Chen, Y. L., Liu, Z. K., Analytis, J. G., Chu, J.-H., Zhang, H. J., Yan, B. H., Mo, S.-K., Moore, R. G., Lu, D. H., Fisher, I. R., Zhang, S. C., Hussain, Z. and Shen, Z.-X. (2010), Single Dirac cone topological surface state and unusual thermoelectric property of compounds from a new topological insulator family, *Phys. Rev. Lett.* **105**, 266401.

33. Brüne, C., Liu, C. X., Novik, E. G., Hankiewicz, E. M., Buhmann, H., Chen, Y. L., Qi, X. L., Shen, Z. X., Zhang, S. C. and Molenkamp, L. W. (2011), Quantum Hall effect from the topological surface states of strained bulk HgTe, *Phys. Rev. Lett.* **106**, 126803.

34. Souma, S., Eto, K., Nomura, M., Nakayama, K., Sato, T., Takahashi, T., Segawa, K. and Ando, Y. (2012), Topological surface states in lead-based ternary telluride $Pb(Bi_{1-x}Sb_x)2Te_4$, *Phys. Rev. Lett.* **108**, 116801.

35. Kuroda, K., Miyahara, H., Ye, M., Eremeev, S. V., Koroteev, Yu. M., Krasovskii, E. E., Chulkov, E. V., Hiramoto, S., Moriyoshi, C., Kuroiwa, Y., Miyamoto, K., Okuda, T., Arita, M., Shimada, K., Namatame, H., Taniguchi, M., Ueda, Y. and Kimura, A. (2012), Experimental verification of $PbBi_2Te_4$ as a 3D topological insulator, *Phys. Rev. Lett.* **108**, 206803.

36. Fu, L. (2009), Hexagonal warping effects in the surface states of the topological insulator Bi_2Te_3, *Phys. Rev. Lett.* **103**, 266801.
37. Liu, C.-X., Qi, X.-L., Zhang, H. J., Dai, X., Fang, Z. and Zhang, S.-C. (2010), Model Hamiltonian for topological insulators, *Phys. Rev. B* **82**, 045122.
38. Alpichshev, Z., Analytis, J. G., Chu, J.-H., Fisher, I. R., Chen, Y. L., Shen, Z. X., Fang, A. and A. Kapitulnik, (2010), STM imaging of electronic waves on the surface of Bi_2Te_3: Topologically protected surface states and hexagonal warping effects, *Phys. Rev. Lett.* **104**, 016401.
39. Kuroda, K., Arita, M., Miyamoto, K., Ye, M., Jiang, J., Kimura, A., Krasovskii, E. E., Chulkov, E. V., Iwasawa, H., Okuda, T., Shimada, K., Ueda, Y., Namatame, H. and Taniguchi, M. (2010), Hexagonally deformed Fermi surface of the 3D topological insulator Bi_2Se_3, *Phys. Rev. Lett.* **105**, 076802.
40. Li, Y.-Y., Wang, G., Zhu, X.-G., Liu, M.-H., Ye, C., Chen, X., Wang, Y.-Y., He, K., Wang, L.-L., Ma, X.-C., Zhang, H.-J., Dai, X., Fang, Z., Xie, X.-C., Liu, Y., Qi, X.-L., Jia, J.-F., Zhang, S.-C. and Xue, Q.-K. (2010), Intrinsic topological insulator Bi_2Te_3 thin films on Si and their thickness limit, *Adv. Mater.* **22**, 4002–4007.
41. Kim, D., Syers, P., Butch, N. P., Paglione, J. and Fuhrer, M. S. (2013), Coherent topological transport on the surface of Bi_2Se_3, *Nat. Commun.* **4**, 2040.
42. Haldane, F. D. M.(1988), Model for a quantum Hall effect without Landau levels: condensed-matter realization of the "parity anomaly", *Phys. Rev. Lett.* **61**, 2015–2018.
43. Jackiw, R. and Rebbi, C. (1976), Solitons with fermion number 1/2, *Phys. Rev. D* **13**, 3398–3409.
44. Berry, M. V. (1984), Quantal phase factors accompanying adiabatic changes, *Proc. Roy. Soc. London* **392**, 45 –57.
45. Thouless, D. J., Kohmoto, M., Nightingale, M. P. and den Nijs, M. (1982), Quantized Hall conductance in a two-dimensional periodic potential, *Phys. Rev. Lett.* **49**, 405–408.
46. Czycholl, G. (2008), *Theoretische Festkörperphysik* (Springer-Verlag, Berlin, Heidelberg).
47. Rammer, J. (2004), *Quantum Transport Theory* (Westview Press, Boulder, CO).
48. Kohmoto, M. (1985), Topological invariant and the quantization of the Hall conductance, *Ann. Phys. (Berlin)* **160**, 343–354.
49. Nielsen, H. B. and Ninomiya, M. (1981), Absence of neutrinos on a lattice: (i) Proof by homotopy theory, *Nucl. Phys. B* **185**, 20–40.

50. Lu, H. Z., Shan, W. Y., Yao, W., Niu, Q. and Shen, S. Q. (2010), Massive Dirac fermions and spin physics in an ultrathin film of topological insulator, *Phys. Rev. B* **81**, 115407.
51. Shan, W. Y., Lu, H. Z. and Shen, S. Q. (2010), Effective continuous model for surface states and thin films of three-dimensional topological insulators, *New J. Phys.* **12**, 043048.
52. Bernevig, B. A. (2013), *Topological insulators and topological superconductors* (Princeton University Press, Princeton, NJ).
53. Chu, J. and Sher, A. (2008), *Physics and Properties of Narrow Gap Semiconductors* (Springer, New York).
54. Büttner, B., Liu, C. X., Tkachov, G., Novik, E. G., Brüne, C., Buhmann, H., Hankiewicz, E. M., Recher, P., Trauzettel, B., Zhang, S. C. and Molenkamp, L. W. (2011), Single valley Dirac fermions in zero-gap HgTe quantum wells, *Nat. Phys.* **7**, 418–422.
55. Culcer, D. (2012), Transport in three-dimensional topological insulators: theory and experiment, *Physica E* **44**, 860.
56. Tkachov, G and Hankiewicz, E. M. (2013), Spin-helical transport in normal and superconducting topological insulators, *Phys. Status Solidi B* **250**, 215.
57. Sheng, L., Sheng, D. N., Ting, C. S. and Haldane, F. D. M. (2005), Nondissipative spin Hall effect via quantized edge transport, *Phys. Rev. Lett.* **95**, 136602.
58. Murakami, S (2006), Quantum spin Hall effect and enhanced magnetic response by spin-orbit coupling, *Phys. Rev. Lett.* **97**, 236805.
59. Zhou, B., Lu, H. Z., Chu, R.-L., Shen, S.-Q. and Niu, Q. (2008), Finite size effects on helical edge states in a quantum spin-Hall system, *Phys. Rev. Lett.* **101**, 246807.
60. Liu, C.-X., Qi, X.-L., Dai, x., Fang, Z. and Zhang, S.-C. (2008), Quantum anomalous Hall effect in $Hg_{1-y}Mn_yTe$ quantum wells, *Phys. Rev. Lett.* **101**, 146802.
61. Tkachov, G. and Hentschel, M. (2009), Spin-orbit coupling, edge states and quantum spin Hall criticality due to Dirac fermion confinement: The case study of graphene, *Eur. Phys. J. B* **69**, 499–504.
62. Yokoyama, T., Tanaka, Y. and Nagaosa, N. (2009), Giant spin rotation in the junction between a normal metal and a quantum spin Hall system, *Phys. Rev. Lett.* **102**, 166801.
63. Schmidt, M. J., Novik, E. G., Kindermann, M. and Trauzettel, B. (2009), Optical manipulation of edge-state transport in HgTe quantum wells in the quantum Hall regime, *Phys. Rev. B* **79**, 241306(R).

64. Akhmerov, A. R., Groth, C. W., Tworzydlo, J. and Beenakker, C. W. J. (2009), Switching of electrical current by spin precession in the first Landau level of an inverted-gap semiconductor, *Phys. Rev. B* **80**, 195320.

65. Groth, C. W., Wimmer, M., Akhmerov, A. R., Tworzydlo, J. and Beenakker, C. W. J. (2009), Theory of the topological Anderson insulator, *Phys. Rev. Lett.* **103**, 196805.

66. Maciejko, J., Qi, X.-L. and Zhang, S.-C. (2010), Magnetoconductance of the quantum spin Hall state, *Phys. Rev. B* **82**, 155310.

67. Tkachov, G. and Hankiewicz, E. M. (2010), Ballistic quantum spin Hall state and enhanced edge backscattering in strong magnetic fields, *Phys. Rev. Lett.* **104**, 166803.

68. Ström, A., Johannesson, H. and Japaridze, G. I. (2010), Edge dynamics in a quantum spin Hall state: effects from Rashba spin–orbit interaction, *Phys. Rev. Lett.* **104**, 256804.

69. Novik, E. G., Recher, P., Hankiewicz, E. M. and Trauzettel, B. (2010), Signatures of topological order in ballistic bulk transport of HgTe quantum wells, *Phys. Rev. B* **81**, 241303(R).

70. Guigou, M. and Cayssol, J. (2010), Andreev spectroscopy of doped HgTe quantum wells, *Phys. Rev. B* **82**, 115312.

71. Adroguer, P., Grenier, C., Carpentier, D., Cayssol, J., Degiovanni, P. and Orignac, E. (2010), Probing the helical edge states of a topological insulator by Cooper-pair injection, *Phys. Rev. B* **82**, 081303.

72. Ostrovsky, P. M., Gornyi, I. V. and Mirlin, A. D. (2010), Interaction-induced criticality in Z_2 topological insulators, *Phys. Rev. Lett.* **105**, 036803.

73. Meidan, D., Micklitz, T. and Brouwer, P.W. (2010), Optimal topological spin pump, *Phys. Rev. B.* **82**, 161303.

74. Rothe, D. G., Reinthaler, R. W., Liu, C.-X., Molenkamp, L. W., Zhang, S.-C. and Hankiewicz, E. M. (2010), Fingerprint of different spin–orbit terms for spin transport in HgTe quantum wells, *New J. Phys.* **12**, 065012.

75. Tkachov, G. and Hankiewicz, E. M. (2011), Transition between ordinary and topological insulator regimes in two-dimensional resonant magnetotransport, *Phys. Rev. B* **83**, 155412.

76. Liu, C.-X., Budich, J. C., Recher, P. and Trauzettel, B. (2011), Charge-spin duality in non-equilibrium transport of helical liquids, *Phys. Rev. B* **83**, 035407.

77. Schmidt, T. L. (2011), Current correlations in quantum spin Hall insulators, *Phys. Rev. Lett.* **107**, 096602.

78. Chang, K. and Lou, W.-K. (2011), Helical quantum states in HgTe quantum dots with inverted band structures, *Phys. Rev. Lett.* **106**, 206802.

79. Michetti, P. and Recher, P. (2011), Bound states and persistent currents in topological insulator rings, *Phys. Rev. B* **83**, 125420.

80. Meidan, D., Micklitz, T. and Brouwer, P.W. (2011), Topological classification of interaction-driven spin pumps, *Phys. Rev. B* **84**, 075325.

81. Virtanen, P. and Recher, P. (2011), Dephasing of spin and charge interference in helical Luttinger liquids, *Phys. Rev. B* **83**, 115332.

82. Wang, Y.-X. and Xiong, S.-J. (2011), Robust transport of quantum spin Hall state in HgTe/CdTe quantum well, *Mod. Phys. Lett. B* **25**, 2001.

83. De Martino, A., Hütten, A. and Egger, R. (2011), Landau levels, edge states, and strained magnetic waveguides in graphene monolayers with enhanced spin-orbit interaction, *Phys. Rev. B* **84**, 155420.

84. Weeks, C., Hu, J., Alicea, J., Franz, M. and Wu, R. (2011), Engineering a robust quantum spin Hall state in graphene via adatom deposition, *Phys. Rev. X* **1**, 021001.

85. Liu, G., Zhou, G. and Chen, Y.-H. (2011), Effect of transverse electric field on helical edge states in a quantum spin-Hall system, *Appl. Phys. Lett.* **99**, 222111.

86. Black-Schaffer, A. M. (2011), Self-consistent superconducting proximity effect at the quantum spin Hall edge, *Phys. Rev. B* **83**, 060504.

87. Timm, C. (2012), Transport through a quantum spin Hall quantum dot, *Phys. Rev. B* **86**, 155456.

88. Krueckl, V. and Richter, K. (2011), Switching spin and charge between edge states in topological insulator constrictions, *Phys. Rev. Lett.* **107**, 086803.

89. Rüegg, A. and Fiete, G. A. (2012), Topological order and semions in a strongly correlated quantum spin-Hall insulator, *Phys. Rev. Lett.* **108**, 046401.

90. Dora, B., Cayssol, J., Simon, F. and Moessner, R. (2012), Optically engineering the topological properties of a spin Hall insulator, *Phys. Rev. Lett.* **108**, 056602.

91. Budich, J. C., Dolcini, F., Recher, P. and Trauzettel, B. (2012), Phonon induced backscattering in helical edge states, *Phys. Rev. Lett.* **108**, 086602.

92. Michetti, P., Budich, J., Novik, E. G. and Recher, P. (2012), Tunable quantum spin Hall effect in double quantum wells, *Phys. Rev. B.* **85**, 125309.

93. Raichev, O. E. (2012), Effective Hamiltonian, energy spectrum, and phase transition induced by in-plane magnetic field in symmetric HgTe quantum wells, *Phys. Rev. B* **85**, 045310.
94. Tkachov, G. and Hankiewicz, E. M. (2012) Two-dimensional topological insulators in quantizing magnetic fields, *Phys. E* **44**, 900.
95. Tkachov, G. and Hentschel, M. (2012), Diffusion on edges of insulating graphene with intravalley and intervalley scattering, *Phys. Rev. B* **86**, 205414.
96. Chen, J.-C., Wang, J. and Sun, Q.-F. (2012), Effect of magnetic field on electron transport in HgTe/CdTe quantum wells: Numerical analysis, *Phys. Rev. B* **85**, 125401.
97. Zou, J. and Jin, G. (2012), $0-\pi$ transition and chiral edge supercurrents in a superconductor/ferromagnetic topological insulator/superconductor junction, *Phys. Rev. B* **85**, 134528.
98. Beugeling, W., Liu, C. X., Novik, E. G., Molenkamp, L. W. and Morais Smith, C. (2012), Reentrant topological phases in Mn-doped HgTe quantum wells, *Phys. Rev. B* **85**, 195304.
99. Goldman, N., Beugeling, W. and Morais Smith, C. (2012), Topological phase transitions between chiral and helical spin textures in a lattice with spin-orbit coupling and a magnetic field, *EPL* **97**, 23003.
100. Kuzmenko, I., Golub, A. and Avishai, Y. (2012), Kondo tunneling into a quantum spin Hall insulator, *Phys. Rev. B* **85**, 205313.
101. Shevtsov, O., Carmier, P., Petitjean, C., Groth, C., Carpentier, D. and Waintal, X. (2012), Graphene-based heterojunction between two topological insulators, *Phys. Rev. X* **2**, 031004.
102. Schmidt, T. L., Rachel, S., von Oppen, F. and Glazman, L. I.(2012), Inelastic electron backscattering in a generic helical edge channel, *Phys. Rev. Lett.* **108**, 156402.
103. Crepin, F., Budich, J. C., Dolcini, F., Recher, P. and Trauzettel, B. (2012), Renormalization group approach for the scattering off a single Rashba impurity in a helical liquid, *Phys. Rev. B* **86**, 121106(R).
104. Scharf, B., Matos-Abiague, A. and Fabian, J. (2012), Magnetic properties of HgTe quantum wells, *Phys. Rev. B* **86**, 075418.
105. Takagaki, Y. (2012), Backscattering from width variations in quasi-one-dimensional strips of topological insulators, *J. Phys.: Condens. Matter* **24** 435301.
106. Hohenadler, M. and Assaad, F. F. (2013), Correlation effects in two-dimensional topological insulators, *J. Phys.: Condens. Matter* **25**, 143201.

107. Budich, J. C., Trauzettel, B. and Sangiovanni, G. (2013), Fluctuation-driven topological Hund insulators, *Phys. Rev. B* **87**, 235104.

108. Brüne, C., Roth, A., Buhmann, H., Hankiewicz, E. M., Molenkamp, L. M., Maciejko, J., Qi, X.-L. and Zhang, S.-C. (2012), Spin polarization of the quantum spin Hall edge states, *Nat. Phys.* **8**, 485–490.

109. König, M., Baenninger, M., Garcia, A. G. F., Harjee, N., Pruitt, B. L., Ames, C., Leubner, P., Brüne, C., Buhmann, H., Molenkamp, L. W. and Goldhaber-Gordon, D. (2013), Spatially resolved study of backscattering in the quantum spin Hall state, *Phys. Rev. X* **3**, 021003.

110. Gusev, G. M., Olshanetsky, E. B., Kvon, Z. D., Levin, A. D., Mikhailov, N. N. and Dvoretsky, S. A. (2012), Nonlocal transport near charge neutrality point in a two-dimensional electron-hole system, *Phys. Rev. Lett.* **108**, 226804.

111. Shuvaev, A. M., Astakhov, G. V., Pimenov, A., Brüne, C., Buhmann, H. and Molenkamp, L. W. (2011), Giant magneto-optical Faraday effect in HgTe thin films in the terahertz spectral range, *Phys. Rev. Lett.* **106**, 107404.

112. Shuvaev, A. M., Astakhov, G. V., Brüne, C., Buhmann, H., Molenkamp, L. W. and Pimenov, A. (2012), Terahertz magneto-optical spectroscopy in HgTe thin films, *Semicond. Sci. Technol.* **27**, 124004.

113. Knez, I., Du, R.-R. and Sullivan, G. (2011), Evidence for helical edge modes in inverted InAs/GaSb quantum wells, *Phys. Rev. Lett.* **107**, 136603.

114. Knez, I. and Du, R.-R. (2012), Quantum spin Hall effect in inverted InAs/GaSb quantum wells, *Frontiers of Physics* **7**, 200–207.

115. Abramowitz, M. and Stegun, I. (1964), *Handbook of Mathematical Functions with Formulas, Graphs, and Mathematical Tables* (National Bureau of Standards, Washington, D.C.).

116. Fisher, D. S. and Lee, P. A. (1981), Relation between conductivity and transmission matrix, *Phys. Rev. B* **23**, 6851–6854.

117. Hancock, J. N., van Mechelen, J. L. M., Kuzmenko, A. B., van der Marel, D., Brüne, C., Novik, E. G., Astakhov, G. V., Buhmann, H. and Molenkamp, L. W. (2011), Surface state charge dynamics of a high-mobility three-dimensional topological insulator, *Phys. Rev. Lett.* **107**, 136803.

118. Shuvaev, A. M., Astakhov, G. V., Tkachov, G., Brüne, C., Buhmann, H., Molenkamp, L. W. and Pimenov, A. (2013), Terahertz quantum Hall effect of Dirac fermions in a topological insulator, *Phys. Rev. B* **87**, 121104(R).

119. Maier, L., Oostinga, J. B., Knott, D., Brüne, C., Virtanen, P., Tkachov, G., Hankiewicz, E. M., Gould, C., Buhmann, H. and Molenkamp,

L. W. (2012), Induced superconductivity in the three-dimensional topological insulator HgTe, *Phys. Rev. Lett.* **109**, 186806.

120. Oostinga, J. B., Maier, L., Schüffelgen, P., Knott, D., Ames, C., Brüne, C., Tkachov, G., Buhmann, H. and Molenkamp, L. W. (2013), Josephson supercurrent through the topological surface states of strained bulk HgTe, *Phys. Rev. X* **3**, 021007.
121. Winkler, R. (2003), *Spin-orbit coupling effects in two-dimensional electron and hole systems* (Springer, Berlin).
122. Tkachov, G., Thienel, C., Pinneker, V., Büttner, B., Brüne, C., Buhmann, H., Molenkamp, L. W. and Hankiewicz, E. M. (2011), Backscattering of Dirac fermions in HgTe quantum wells with a finite gap, *Phys. Rev. Lett.* **106**, 076802.
123. Altshuler, B. L., Khmel'nitzkii, D., Larkin, A. I. and Lee, P. A. (1980), Magnetoresistance and Hall effect in a disordered two-dimensional electron gas, *Phys. Rev. B* **22**, 5142–5153.
124. Bergmann, G. (1984), Weak localization in thin films: A time-of-flight experiment with conduction electrons, *Phys. Rep.* **107**, 1–58.
125. Akkermans, E. and Montambaux, G. (2007), *Mesoscopic Physics of Electrons and Photons* (Cambridge University Press, Cambridge).
126. Hikami, S., Larkin, A. I. and Nagaoka, Y. (1980), Spin–orbit interaction and magnetoresistance in the two dimensional random system, *Prog. Theor. Phys.* **63**, 707–710.
127. Iordanskii, S. V., Lyanda-Geller, Yu. B. and Pikus, G. E. (1994), Weak localization in quantum wells with spin–orbit interaction, *JETP Lett.* **60**, 206.
128. Knap, W., Skierbiszewski, C., Zduniak, A., Litwin-Staszewska, E., Bertho, D., Kobbi, F., Robert, J. L., Pikus, G. E., Pikus, F. G., Iordanskii, S. V., Mosser, V., Zekentes, K. and Lyanda-Geller, Yu. B. (1996), Weak antilocalization and spin precession in quantum wells, *Phys. Rev. B* **53**, 3912–3924.
129. Aleiner, I. L. and Falko, V. I. (2001), Spin–orbit coupling effects on quantum transport in lateral semiconductor dots, *Phys. Rev. Lett.* **87**, 256801.
130. Meyer, J. S., Altland, A. and Altshuler, B. L. (2002), Quantum transport in parallel magnetic fields: A realization of the Berry–Robnik symmetry phenomenon, *Phys. Rev. Lett.* **89**, 206601.
131. Miller, J. B., Zumbühl, D. M., Marcus, C. M., Lyanda-Geller, Y. B., Goldhaber-Gordon, D., Campman, K. and Gossard, A. C. (2003), Gate-controlled spin–orbit quantum interference effects in lateral transport, *Phys. Rev. Lett.* **90**, 076807.

132. Schäpers, Th., Guzenko, V. A., Pala, M. G., Zülicke, U., Governale, M., Knobbe, J. and Hardtdegen, H. (2006), Suppression of weak antilocalization in $Ga_xIn_{1-x}As/InP$ narrow quantum wires, *Phys. Rev. B* **74**, 081301(R).
133. Neumaier, D., Wagner, K., Geißler, S., Wurstbauer, U., Sadowski, J., Wegscheider, W. and Weiss, D. (2007), Weak localization in ferromagnetic (Ga,Mn)As nanostructures, *Phys. Rev. Lett.* **99**, 116803.
134. Olshanetsky, E. B., Kvon, Z. D., Gusev, G. M., Mikhailov, N. N., Dvoretsky, S. A. and Portal, J. C. (2010), Weak antilocalization in HgTe quantum wells near a topological transition, *JETP Lett.* **91**, 347–350.
135. Suzuura, H. and Ando, T. (2002), Crossover from symplectic to orthogonal class in a two-dimensional honeycomb lattice, *Phys. Rev. Lett.* **89**, 266603.
136. Ando, T. (2005), Theory of electronic states and transport in carbon nanotubes, *J. Phys. Soc. Jpn.* **74**, 777.
137. Morozov, S.V., Novoselov, K. S., Katsnelson, M. I., Schedin, F., Ponomarenko, L. A., Jiang, D. and Geim, A. K. (2006), Strong suppression of weak localization in graphene, *Phys. Rev. Lett.* **97**, 016801.
138. McCann, E., Kechedzhi, K., Fal'ko, V.I., Suzuura, H., Ando, T. and Altshuler, B. L. (2006), Weak-localization magnetoresistance and valley symmetry in graphene, *Phys. Rev. Lett.* **97**, 146805.
139. Tikhonenko, F. V., Kozikov, A. A., Savchenko, A. K. and Gorbachev, R. V. (2009), Transition between electron localization and antilocalization in graphene, *Phys. Rev. Lett.* **103**, 226801.
140. Imura, K.-I., Kuramoto, Y. and Nomura, K. (2010), Anti-localization of graphene under the substrate electric field, *Europhys. Lett.* **89**, 17009.
141. Wurm, J., Rycerz, A., Adagideli, I., Wimmer, M., Richter, K. and Baranger, H. U. (2009), Symmetry classes in graphene quantum dots: Universal spectral statistics, weak localization, and conductance fluctuations, *Phys. Rev. Lett.* **102**, 056806.
142. Ortmann, F., Cresti, A., Montambaux, G. and Roche, S. (2011), Magnetoresistance in disordered graphene: The role of pseudospin and dimensionality effects unraveled, *Europhys. Lett.* **94**, 47006.
143. Minke, S., Bundesmann, J., Weiss, D. and Eroms, J. (2012), Phase coherent transport in graphene nanoribbons and graphene nanoribbon arrays, *Phys. Rev. B* **86**, 155403.
144. Checkelsky, J. G., Hor, Y. S., Liu, M. H., Qu, D. X., Cava, R. J. and Ong, N. P. (2009), Quantum interference in macroscopic crystals of nonmetallic Bi_2Se_3, *Phys. Rev. Lett.* **103**, 246601.

145. Chen, J., Qin, H. J., Yang, F., Liu, J., Guan, T., Qu, F. M., Zhang, G. H., Shi, J. R., Xie, X. C., Yang, C. L., Wu, K. H., Li, Y. Q. and Lu, L. (2010), Gate-voltage control of chemical potential and weak antilocalization in Bi_2Se_3, *Phys. Rev. Lett.* **105**, 176602.

146. He, H.-T., Wang, G., Zhang, T., Sou, I.-K., Wong, G. K. L., Wang, J.-N., Lu, H.-Z., Shen, S.-Q. and Zhang, F.-C. (2011), Impurity effect on weak antilocalization in the topological insulator Bi_2Te_3, *Phys. Rev. Lett.* **106**, 166805.

147. Lu, H.-Z., Shi, J. and Shen, S.-Q. (2011), Competition between weak localization and antilocalization in topological surface states, *Phys. Rev. Lett.* **107**, 076801.

148. Tkachov, G. and Hankiewicz, E. M. (2011), Weak antilocalization in HgTe quantum wells and topological surface states: Massive versus massless Dirac fermions, *Phys. Rev. B* **84**, 035444.

149. Nestoklon, M. O., Averkiev, N. S. and Tarasenko, S. A. (2011), Weak localization of two-dimensional Dirac fermions beyond the diffusion regime, *Sol. State Comm.* **151**, 1550–1553.

150. Lu, H.-Z. and Shen, S.-Q. (2011), Weak localization of bulk channels in topological insulator thin films, *Phys. Rev. B* **84**, 125138.

151. Steinberg, H., Laloe, J.-B., Fatemi, V., Moodera, J. S. and Jarillo-Herrero, P. (2011), Electrically tunable surface-to-bulk coherent coupling in topological insulator thin films, *Phys. Rev. B* **84**, 233101.

152. Kim, Y. S., Brahlek, M., Bansal, N., Edrey, E., Kapilevich, G. A., Iida, K., Tanimura, M., Horibe, Y., Cheong, S.-W. and Oh, S. (2011), Thickness-dependent bulk properties and weak antilocalization effect in topological insulator Bi_2Se_3, *Phys. Rev. B* **84**, 073109.

153. Chen, J., He, X. Y., Wu, K. H., Ji, Z. Q., Lu, L., Shi, J. R., Smet, J. H. and Li, Y. Q. (2011), Tunable surface conductivity in Bi_2Se_3 revealed in diffusive electron transport, *Phys. Rev. B* **83**, 241304.

154. Minkov, G. M., Germanenko, A. V., Rut, O. E., Sherstobitov, A. A., Dvoretski, S. A. and Mikhailov, N. N. (2012), Weak antilocalization in HgTe quantum wells with inverted energy spectra, *Phys. Rev. B* **85**, 235312.

155. Cha, J. J., Kong, D., Hong, S.-S., Analytis, J. G., Lai, K. and Cui, Y. (2012), Weak antilocalization in $Bi_2(Se_xTe_{1-x})_3$ nanoribbons and nanoplates, *Nano Lett.* **12**, 1107–1111.

156. Takagaki, Y., Jenichen, B., Jahn, U., Ramsteiner, M. and Friedland, K.-J. (2012), Weak antilocalization and electron–electron interaction effects in Cu-doped Bi_2Se_3 films, *Phys. Rev. B* **85**, 115314.

157. Takagaki, Y., Giussani, A., Perumal, K., Calarco, R. and Friedland, K.-J. (2012), Robust topological surface states in Sb_2Te_3 layers as seen from the weak antilocalization effect, *Phys. Rev. B* **86**, 125137.

158. Krueckl, V. and Richter, K. (2012), Probing the band topology of mercury telluride through weak localization and antilocalization, *Semicond. Sci. Technol.* **27**, 124006.

159. Ostrovsky, P. M., Gornyi, I. V. and Mirlin, A. D. (2012), Symmetries and weak localization and antilocalization of Dirac fermions in HgTe quantum wells, *Phys. Rev. B* **86**, 125323.

160. Adroguer, P., Carpentier, D., Cayssol, J. and Orignac, E. (2012), Diffusion at the surface of topological insulators, *New J. Phys.* **14**, 103027.

161. Garate, I. and Glazman, L. (2012), Weak localization and antilocalization in topological insulator thin films with coherent bulk-surface coupling, *Phys. Rev. B* **86**, 035422.

162. Tian, C. (2012), Anomalous quantum diffusion and the topological metal, *Phys. Rev. B* **86**, 121304(R).

163. Takagaki, Y., Giussani, A., Tominaga, J., Jahn, U. and Calarco, R. (2013), Transport properties in a Sb–Te binary topological-insulator system, *J. Phys.: Condens. Matter* **25** 345801.

164. Bychkov, Y. A. and Rashba, E. I. (1984), Oscillatory effects and the magnetic susceptibility of carriers in inversion layers, *J. Phys. C, Solid State Phys.* **17**, 6039.

165. Zutic, I., Fabian, J. and Das Sarma, S. (2004), Spintronics: Fundamentals and applications, *Rev. Mod. Phys.* **76**, 323–410.

166. Pesin, D. and MacDonald, A. H. (2012), Spintronics and pseudospintronics in graphene and topological insulators, *Nature Mat.* **11**, 409–416.

167. Taskin, A. A., Segawa, K. and Ando, Y. (2010), Oscillatory angular dependence of the magnetoresistance in a topological insulator $Bi_{1-x}Sb_x$, *Phys. Rev. B* **82**, 121302(R).

168. Butch, N., Kirshenbaum, K., Syers, P., Sushkov, A., Jenkins, G., Drew, H. and Paglione, J. (2010), Strong surface scattering in ultrahigh-mobility Bi_2Se_3 topological insulator crystals, *Phys. Rev. B* **81**, 241301(R).

169. Eto, K., Ren, Z., Taskın, A., Segawa, K. and Ando, Y. (2010), Angular-dependent oscillations of the magnetoresistance in Bi_2Se_3 due to the three-dimensional bulk Fermi surface, *Phys. Rev. B* **81**, 195309.

170. Analytis, J., Chu, J., Chen, Y., Corredor, F., McDonald, R., Shen, Z. and Fisher, I. (2010), Bulk Fermi surface coexistence with Dirac surface

state in Bi_2Se_3: A comparison of photoemission and Shubnikov–de Haas measurements, *Phys. Rev. B* **81**, 205407.

171. LaForge, A. D., Frenzel, A., Pursley, B. C., Lin, T., Liu, X., Shi, J. and Basov, D. N. (2010), Optical characterization of Bi_2Se_3 in a magnetic field: Infrared evidence for magnetoelectric coupling in a topological insulator material, *Phys. Rev. B* **81**, 125120.
172. Sushkov, A. B., Jenkins, G. S., Schmadel, D. C., Butch, N. P., Paglione, J. and Drew, H. D. (2010), Far-infrared cyclotron resonance and Faraday effect in Bi_2Se_3, *Phys. Rev. B* **82**, 125110.
173. Taskin, A. A., Ren, Z., Sasaki, S., Segawa, K. and Ando, Y. (2011), Observation of Dirac holes and electrons in a topological insulator, *Phys. Rev. Lett.* **107**, 016801.
174. Valdes Aguilar, R., Stier, A. V., Liu, W., Bilbro, L. S., George, D. K., Bansal, N., Wu, L., Cerne, J., Markelz, A. G., Oh, S. and Armitage, N. P. (2012), Terahertz response and Colossal Kerr rotation from the surface states of the topological insulator Bi_2Se_3, *Phys. Rev. Lett.* **108**, 087403.
175. Qi, X.-L., Hughes, T. L. and Zhang, S.-C. (2008), Topological field theory of time-reversal invariant insulators, *Phys. Rev. B* **78**, 195424.
176. Essin, A. M., Moore, J. E. and Vanderbilt, D. (2009), Magnetoelectric polarizability and axion electrodynamics in crystalline insulators, *Phys. Rev. Lett.* **102**, 146805.
177. Mondal, S., Sen, D., Sengupta, K. and Shankar, R. (2010), Magnetotransport of Dirac fermions on the surface of a topological insulator, *Phys. Rev. B* **82**, 045120.
178. Tse, W.-K. and MacDonald, A. H. (2010), Giant magneto-optical Kerr effect and universal Faraday effect in thin-film topological insulators, *Phys. Rev. Lett.* **105**, 057401.
179. Wu, Z., Peeters, F. M. and Chang, K. (2010), Electron tunneling through double magnetic barriers on the surface of a topological insulator, *Phys. Rev. B* **82**, 115211.
180. Tse W.-K. and MacDonald, A. H. (2010), Magneto-optical and magnetoelectric effects of topological insulators in quantizing magnetic fields, *Phys. Rev. B* **82**, 161104(R).
181. Maciejko, J., Qi, X.-L., Drew, H. D. and Zhang, S.-C. (2010), Topological quantization in units of the fine structure constant, *Phys. Rev. Lett.* **105**, 166803.
182. Zazunov, A., Kundu, A., Hütten, A. and Egger, R. (2010), Magnetic scattering of Dirac fermions in topological insulators and graphene, *Phys. Rev. B* **82**, 155431.

183. Culcer, D., Hwang, E. H., Stanescu, T. D. and Das Sarma, S. (2010), Two-dimensional surface charge transport in topological insulators, *Phys. Rev. B* **82**, 155457.

184. Bardarson, J. H., Brouwer, P. W. and Moore, J. E. (2010), Aharonov–Bohm oscillations in disordered topological insulator nanowires, *Phys. Rev. Lett.* **105**, 156803.

185. Yazyev, O. V., Moore, J. E. and Louie, S. G. (2010), Spin polarization and transport of surface states in the topological insulators Bi_2Se_3 and Bi_2Te_3 from first principles, *Phys. Rev. Lett.* **105**, 266806.

186. Chu, R.-L., Shan, W.-Y., Lu, J. and Shen, S.-Q. (2011), Surface and edge states in topological semimetals, *Phys. Rev. B* **83**, 075110.

187. Zhu, J.-J., Yao, D.-X., Zhang, S.-C. and Chang, K. (2011), Electrically controllable surface magnetism on the surface of topological insulators, *Phys. Rev. Lett.* **106**, 097201.

188. Li, Q., Ghosh, P., Sau, J. D., Tewari, S. and Das Sarma, S. (2011), Anisotropic surface transport in topological insulators in proximity to a helical spin density wave, *Phys. Rev. B* **83**, 085110.

189. Dora, B. and Moessner, R. (2011), Dynamics of the spin Hall effect in topological insulators and graphene, *Phys. Rev. B* **83**, 073403.

190. Zyuzin, A. A. and Burkov, A. A. (2011), Thin topological insulator film in a perpendicular magnetic field, *Phys. Rev. B* **83**, 195413.

191. Culcer, D. and Das Sarma, S. (2011), Anomalous Hall response of topological insulators, *Phys. Rev. B* **83**, 245441.

192. Tkachov, G. and Hankiewicz, E. M. (2011), Anomalous galvano-magnetism, cyclotron resonance, and microwave spectroscopy of topological insulators, *Phys. Rev. B* **84**, 035405.

193. Burkov, A. A. and Balents, L. (2011), Weyl semimetal in a topological insulator multilayer, *Phys. Rev. Lett.* **107**, 127205.

194. Wang, C. M. and Yu, F. J. (2011), Effects of hexagonal warping on surface transport in topological insulators, *Phys. Rev. B* **84**, 155440.

195. Tse, W.-K. and MacDonald, A. H. (2011), Magneto-optical Faraday and Kerr effects in topological insulator films and in other layered quantized Hall systems, *Phys. Rev. B* **84**, 205327.

196. Culcer, D. (2011), Linear response theory of interacting topological insulators, *Phys. Rev. B* **84**, 235411.

197. Papic, Z., Abanin, D. A., Barlas, Y. and Bhatt, R. N. (2011), Tunable interactions and phase transitions in Dirac materials in a magnetic field, *Phys. Rev. B* **84**, 241306(R).

198. Fu, Z.-G., Zhang, P. and Li, S.-S. (2011), Aharonov–Bohm oscillations in the local density of topological surface states, *Appl. Phys. Lett.* **99**, 243110.

199. Tserkovnyak, Y. and Loss, D. (2012), Thin-film magnetization dynamics on the surface of a topological insulator, *Phys. Rev. Lett.* **108**, 187201.

200. Abanin, D. A., Papic, Z., Barlas, Y. and Bhatt, R. N. (2012), Stability of the k=3 Read-Rezayi state in chiral two-dimensional systems with tunable interactions, *New J. Phys.* **14**, 025009.

201. Bergman, D. L. and Refael, G. (2013), Thermodynamic measure of the magneto-electric coupling in a three-dimensional topological insulator, *Phys. Rev. B* **87**, 024202.

202. Tudorovskiy, T. and Katsnelson, M. I. (2012), Peculiarities of dynamics of Dirac fermions associated with zero-mass lines, *Phys. Rev. B* **86**, 045419.

203. Tahir, M. and Schwingenschlögl, U. (2012), Quantum magnetotransport properties of topological insulators under strain, *Phys. Rev. B* **86**, 075310.

204. She, J.-H., Fransson, J., Bishop, A. R. and Balatsky, A. V. (2013), Inelastic electron tunneling spectroscopy for topological insulators, *Phys. Rev. Lett.* **110**, 026802.

205. Bardarson, J. H. and Moore, J. E. (2013), Quantum interference and Aharonov–Bohm oscillations in topological insulators, *Rep. Prog. Phys.* **76**, 056501.

206. Bardeen, J., Cooper, L. N. and Schrieffer, J. R. (1957), Theory of Superconductivity, *Phys. Rev.* **108**, 1175–1204.

207. Gorkov, L.P. (1958), On the energy spectrum of superconductors, *J. Exp. Theor. Phys.* **34**, 735–739; [Sov. Phys. JETP **7**, 505–508].

208. de Gennes, P. G. (1964), Boundary Effects in Superconductors, *Rev. Mod. Phys.* **36**, 225–237.

209. Fu, L. and Kane, C. L. (2008), Superconducting proximity effect and Majorana fermions at the surface of a topological insulator, *Phys. Rev. Lett.* **100**, 096407.

210. Stanescu, T. D., Sau, J. D., Lutchyn, R. M. and Das Sarma, S. (2010), Proximity effect at the superconductor–topological insulator interface, *Phys. Rev. B* **81**, 241310(R).

211. Potter, A. C. and Lee, P. A. (2011), Engineering a $p+ip$ superconductor: Comparison of topological insulator and Rashba spin–orbit-coupled materials, *Phys. Rev. B* **83**, 184520.

212. Lababidi, M. and Zhao, E. (2011), Microscopic simulation of superconductor–topological insulator proximity structures, *Phys. Rev. B* **83**, 184511.

213. Ito, Y., Yamaji, Y. and Imada, M. (2011), Stability of unconventional superconductivity on surfaces of topological insulators, *J. Phys. Soc. Jpn.* **80**, 063704.

214. Khaymovich, I. M., Chtchelkatchev, N. M. and Vinokur, V. M. (2011), Instability of topological order and localization of edge states in HgTe quantum wells coupled to s-wave superconductor, *Phys. Rev. B* **84**, 075142.

215. Virtanen, P. and Recher, P. (2012), Signatures of Rashba spin–orbit interaction in the superconducting proximity effect in helical Luttinger liquids, *Phys. Rev. B* **85**, 035310.

216. Yokoyama, T., (2012), Josephson and proximity effects on the surface of a topological insulator, *Phys. Rev. B* **86**, 075410.

217. Black-Schaffer, A. M. and Balatsky, A. V. (2012), Odd-frequency superconducting pairing in topological insulators, *Phys. Rev. B* **86**, 144506.

218. Tkachov, G. (2013), Suppression of surface *p*-wave superconductivity in disordered topological insulators, *Phys. Rev. B* **87**, 245422.

219. Zhang, D., Wang, J., DaSilva, A. M., Lee, J. S., Gutierrez, H. R., Chan, M. H. W., Jain, J. and Samarth, N. (2011), Superconducting proximity effect and possible evidence for Pearl vortices in a candidate topological insulator, *Phys. Rev. B* **84**, 165120.

220. Koren, G., Kirzhner, T., Lahoud, E., Chashka, K. B. and Kanigel, A. (2011), Proximity-induced superconductivity in topological Bi_2Te_2Se and Bi_2Se_3 films: Robust zero-energy bound state possibly due to Majorana fermions, *Phys. Rev. B* **84**, 224521.

221. Sacepe, B., Oostinga, J. B., Li, J., Ubaldini, A., Couto, N. J. G., Giannini, E. and Morpurgo, A. F. (2011), Gate-tuned normal and superconducting transport at the surface of a topological insulator, *Nat. Commun.* **2** 575.

222. Veldhorst, M., Snelder, M., Hoek, M., Gang, T., Guduru, V. K., Wang, X. L., Zeitler, U., Wiel, W. G. v. d., Golubov, A. A. and Hilgenkamp, H. (2012), Josephson supercurrent through a topological insulator surface state, *Nature Mat.* **11**, 417–421.

223. Wang, M.-X., Liu, C., Xu, J.-P., Yang, F., Miao, L., Yao, M.-Y., Gao, C. L., Shen, C., Ma, X., Chen, X., Xu, Z.-A., Liu, Y., Zhang, S.-C., Qian, D., Jia, J.-F. and Xue, Q.-K. (2012), The coexistence of superconductivity and topological order in the Bi_2Se_3 thin films, *Science* **336**, 52–55.

224. Williams, J. R., Bestwick, A. J., Gallagher, P., Hong, S. S., Cui, Y., Bleich, A. S., Analytis, J. G., Fisher, I. R. and Goldhaber-Gordon, D. (2012), Unconventional Josephson effect in hybrid superconductor-topological insulator devices, *Phys. Rev. Lett.* **109**, 056803.
225. Cho, S., Dellabetta, B., Yang, A., Schneeloch, J., Xu, Z., Valla, T., Gu, G., Gilbert, M. J. and Mason, N. (2013), Symmetry protected Josephson supercurrents in three-dimensional topological insulators, *Nat. Commun.* **4**, 1689.
226. Snelder, M., Veldhorst, M., Golubov, A. A. and Brinkman, A., (2013), Andreev bound states and current-phase relations in three-dimensional topological insulators, *Phys. Rev. B* **87**, 104507.
227. Tkachov, G. and Hankiewicz, E. M. (2013), Helical Andreev bound states and superconducting Klein tunneling in topological insulator Josephson junctions, *Phys. Rev. B* **88**, 075401.
228. Beenakker, C. W. J. (1992). Three "universal" mesoscopic Josephson effects, in *Transport Phenomena in Mesoscopic Systems*, edited by H. Fukuyama and T. Ando (Springer, Berlin), pp. 235–253.
229. Olund, C. T. and Zhao, E. (2012), Current-phase relation for Josephson effect through helical metal, *Phys. Rev. B* **86**, 214515.
230. Fu, L. and Kane, C. L. (2009), Josephson current and noise at a superconductor-quantum spin Hall insulator-superconductor junction, *Phys. Rev. B* **79**, 161408(R).
231. Akhmerov, A. R., Nilsson, J. and Beenakker, C. W. J. (2009), Electrically detected interferometry of Majorana fermions in a topological insulator, *Phys. Rev. Lett.* **102**, 216404.
232. Law, K. T., Lee, P. A. and Ng, T. K. (2009), Majorana fermion induced resonant Andreev reflection, *Phys. Rev. Lett.* **103**, 237001.
233. Tanaka, Y., Yokoyama, T. and Nagaosa, N. (2009), Manipulation of the Majorana fermion, Andreev reflection, and Josephson current on topological insulators, *Phys. Rev. Lett.* **103**, 107002.
234. Linder, J., Tanaka, Y., Yokoyama, T., Sudbo, A. and Nagaosa, N. (2010), Interplay between superconductivity and ferromagnetism on a topological insulator, *Phys. Rev. B* **81**, 184525.
235. Sau, J. D., Lutchyn, R. M., Tewari, S. and Das Sarma, S. (2010), Robustness of Majorana fermions in proximity-induced superconductors, *Phys. Rev. B* **82**, 094522.
236. Badiane, D. M., Houzet, M. and Meyer, J. S. (2011), Non-equilibrium Josephson effect through helical edge states, *Phys. Rev. Lett.* **107**, 177002.

237. Ioselevich, P. A. and Feigelman, M. V. (2011), Anomalous Josephson current via Majorana bound states in topological insulators, *Phys. Rev. Lett.* **106**, 077003.
238. Golub, A., Kuzmenko, I. and Avishai, Y. (2011), Kondo correlations and Majorana bound states in a metal to quantum-dot to topological-superconductor junction, *Phys. Rev. Lett.* **107**, 176802.
239. Yamakage, A., Sato, M., Yada, K., Kashiwaya, S. and Tanaka, Y. (2013), Anomalous Josephson current in superconducting topological insulator, *Phys. Rev. B* **87**, 100510(R).
240. Beenakker, C. W. J., Pikulin, D. I., Hyart, T., Schomerus, H. and Dahlhaus, J. P. (2013), Fermion-parity anomaly of the critical supercurrent in the quantum spin-Hall effect, *Phys. Rev. Lett.* **110**, 017003.
241. Potter, A. C. and Fu, L. (2013), Anomalous supercurrent from Majorana states in topological insulator Josephson junctions, *Phys. Rev. B* **88**, 121109(R).
242. Oreg, Y., Refael, G. and von Oppen, F. (2010), Helical liquids and majorana bound states in quantum wires, *Phys. Rev. Lett.* **105**, 177002.
243. Sau, J. D., Lutchyn, R. M., Tewari, S. and Das Sarma, S. (2010), Generic new platform for topological quantum computation using semiconductor heterostructures, *Phys. Rev. Lett.* **104**, 040502.
244. Lutchyn, R. M., Sau, J. D. and Das Sarma, S. (2010), Majorana fermions and a topological phase transition in semiconductor–superconductor heterostructures, *Phys. Rev. Lett.* **105**, 077001.
245. Mourik, V., Zuo, K., Frolov, S. M., Plissard, S. R., Bakkers, E. P. A. M. and Kouwenhoven, L. P. (2012), Signatures of Majorana fermions in hybrid superconductor-semiconductor nanowire devices, *Science* **336**, 1003.
246. Alicea, J. (2012), New directions in the pursuit of Majorana fermions in solid state systems, *Rep. Prog. Phys.* **75**, 076501.
247. Beenakker, C.W.J. (2013), Search for Majorana fermions in superconductors, *Annu. Rev. Con. Mat. Phys.* **4**, 113–136.

Index

For Product Safety Concerns and Information please contact our EU
representative GPSR@taylorandfrancis.com
Taylor & Francis Verlag GmbH, Kaufingerstraße 24, 80331 München, Germany

www.ingramcontent.com/pod-product-compliance
Ingram Content Group UK Ltd.
Pitfield, Milton Keynes, MK11 3LW, UK
UKHW022201190926
475702UK00003B/150
* 9 7 8 9 8 1 4 6 1 3 2 5 5 *